About the author

Dr Mark Everard's work in all four sectors of society – private, public, academic and voluntary – has taken him across five continents to undertake applied research, policy development and capacity-building relating to the ways in which people connect with ecosystems. The author of 14 other books, including *Common Ground* (Zed Books, 2011) and *The Hydropolitics of Dams* (Zed Books, 2013), over 60 peer-reviewed scientific papers and over 250 technical magazine articles, Mark is also a communicator on sustainability and wider environmental and resource use matters on TV and radio. He has served on numerous government advisory and expert groups in the UK, as well as advising other governments and multinational corporations on sustainability matters. His speciality is systemic thinking, particularly around connections between the water environment and other environmental media and the human activities that depend on and influence them. Mark's work includes environmental ethics and economics as a means to bring our intimate interdependencies with ecosystems into the mainstream of public awareness and government thinking.

Breathing space

The natural and unnatural history of air

MARK EVERARD

Zed Books
LONDON

Breathing Space: The natural and unnatural history of air was first published in 2015 by Zed Books Ltd, 7 Cynthia Street, London N1 9JF, UK
www.zedbooks.co.uk

Set in Arnhem Pro Blond and Futura by Ewan Smith, London
Index: ed.emery@thefreeuniversity.net
Cover designed by www.roguefour.co.uk
Cover image © Getty/alexihobbs

A catalogue record for this book is available from the British Library

ISBN 978-1-78360-385-5 hb
ISBN 978-1-78360-384-8 pb
ISBN 978-1-78360-386-2 pdf
ISBN 978-1-78360-387-9 epub
ISBN 978-1-78360-388-6 mobi

Printed in the USA by Edwards Brothers Malloy

Contents

Acknowledgements

The author would like to thank his colleagues at the University of the West of England – Ben Pontin, Tom Appleby, Chad Staddon, Enda Hayes, Jo Barnes and Jim Longhurst – with whom he worked to co-produce the paper 'Air as a common good', published in 2013 in the scientific journal *Environmental Science and Policy* (volume 33, pp. 354–68). He would also like to thank his long-suffering family, Jackie and Daisy!

Introduction

The air and the atmosphere, respectively the physical 'stuff' we breathe and the wider gaseous and energetic layers surrounding our home planet, comprise immensely complex and dynamic flows of energy, matter and living things. The atmosphere is, by orders of magnitude, the largest habitat on our home planet. Yet, due to its fluidity, invisibility and lack of clear national ownership, it is also the most overlooked. We must not allow the continuing omission of the atmosphere as an integrated whole from our thinking and development plans, as this leads to responses that are fragmented and that often appear only when acute problems manifest themselves, and ignores impacts elsewhere in the air system.

It is an irony that, although dry air is an efficient electrical insulator, in so many other respects the atmosphere is the planet's great connector. The air transmits light, allows the conduction, convection and radiation of heat, and distributes moisture and other chemical substances throughout the globe. Air is the medium of distribution of scent trails and pheromones, of spores, pollen and seeds, and of disease organisms and their vectors. Climate systems redistribute energy captured from the sun, retained in ever greater quantities by the global greenhouse yet countervailed by the albedo effect of cloud cover, and with it evaporation from the seas and other moist surfaces. Birds, insects and other organisms fly in it, and these creatures, the pleasant vista of the sky and the sense of space and 'fresh air' relax and renew us.

Air touches and is intimately interdependent with the earth and water, shaping, conveying and otherwise interacting with their constituents. The atmosphere has evolved since our home planet was formed from the condensation of dust in the embryonic solar system, and it has been profoundly shaped by the emergence and activities of life. Indeed, the very nature of the atmosphere carries the fingerprints of the living processes that form and maintain it, the whole living biosphere of interacting non-living and living elements recycling, reforming and regulating itself as a single 'superorganism' that some refer to as 'Gaia'.

Of course, these interdependent life forms include humanity. Our diverse biological needs, economic and subsistence activities and wider life enjoyment depend utterly on the protective and sustaining properties of the thin gaseous skin surrounding our home planet. And, equally, all of the activities of our technologically advanced and populous species have impacts, often profound, upon the airspace and all who share it. Indeed, what makes life possible, profitable, enjoyable and sustainable is the regeneration and cleansing of air, the watering through atmospheric circulation systems of the land that grows our food, the recharge of rivers, lakes and groundwater, and the flows of energy through the atmosphere that we harvest in turbines and rely on to bear away waste gases, or that we exploit in other ways. We use the air to send messages of peace both acoustically and electronically, but also to bear chemical weapons, launch bombing raids and secure aerial dominance in warfare and militarily enforced peace-making.

Ask any scuba diver and they will tell you interesting things about air: how important it is to ensure that your kit is suitable and well serviced so that it will not fail when you need it most; that alternative air sources are an essential part of the scuba kit; and, importantly, that the first thing to learn is how to share air with a dive buddy. Thirty metres down deep, there is no room for error. Without air, at any depth or altitude, we die. But sharing air is a matter of far greater significance than just for water sports. As shareholders in and dependants of by far the largest habitat of this planet, we share air not merely with our buddies but with all of global civilisation and with every life form. Acting as if we were all connected is therefore something that should come naturally; as we have seen, bad things can happen if we disregard the air's importance or treat it in a fragmented manner. Air, then, is the ultimate common for all life, including prospects for all of humanity. It is the medium through which we visit inequities upon others, particularly those yet to inherit it. The air is the medium that joins us and, above all, a key element of nature's integral interconnectedness.

Breathing Space: The natural and unnatural history of air takes a fresh look at the ocean of gases and physical shields immersing and surrounding us every moment of our lives. It examines its evolution and characteristics as well as the benefits that flow from the atmosphere to us and the harm that we inadvertently inflict upon it. We explore responses to some acute problems wrought upon the air and upper atmosphere at our hand, before then questioning whether our vision and ensuing decisions and actions are wide-ranging enough to safeguard one of the most fundamental of natural resources underpinning our collective future.

Breathing Space is not a book for the specialist in any one aspect of the atmosphere, be that local air quality management, ozone depletion or climate change. Those topics, and many more, are considered in some detail. However, the principal emphasis is on the whole integrated air–human system, and on the lessons that can emerge from looking at our systemic interdependence. It is about air as an ecosystem, as a common resource vital to all people now and to those yet to come. It is about what we need to do and how we might do it if we are to safeguard our common inheritance and common legacy.

Chapter 1: *Air and the making of the atmosphere*, highlights the properties of the airspace and notes why these have contributed to its widespread omission from major global, national and local studies, such as the United Nations Millennium Ecosystem Assessment and the UK National Ecosystem Assessment. The formation and subsequent evolution of the atmosphere are considered, as is its current structure and the major properties of atmospheric layers. The depth and significance of the interdependence of the atmospheric system with landscapes and life are summarised, along with key properties and the functioning of the climatic system and weather. We consider the deep interactions between life and the environment that supports it: the dynamic biosphere of our home planet.

Chapter 2: *Living in a bubble*, relates to the biological interdependence of people with air and the atmospheric system. It addresses how we are biologically embedded in the bubble of air that surrounds planet Earth, touching on the role of the atmosphere in human senses, the many ways in which our deep relationship with air and the atmosphere shapes beliefs and myths, the places where we live and the livelihoods within them, and how the air influences our technological development and aesthetic expression.

Chapter 3: *What does air do for us?*, then considers the breadth of benefits we derive from air and the atmosphere within the context of the ecosystem services framework. It explores prior studies and new insights from discussions throughout this book to draw out these benefits by ecosystem service category and by atmospheric layer.

Chapter 4: *Abuses of the air*, outlines the many ways in which human activities have impacted on aspects of both air and the atmosphere. It considers how all matter entering the common airspace spreads and may accumulate with unforeseen consequences. Our industrial trajectory is discussed, as are broader implications for air and the atmosphere arising from the massive conversion of terrestrial habitats. Forests and wetlands in particular are addressed due to their deep influence on the regeneration and regulation of the airspace and its services. Significant threats such as ozone depletion and climate change are highlighted, together with a range of emerging concerns.

Chapter 5: *Managing our impacts on air*, explores the timelines of various forms of legislative and other responses to address pressures on the air and the atmosphere. Examples of our responses include how we have tackled key challenges such as acid rain, ozone depletion and climate change, and some of the lessons that arise from this. The chapter explores the evolution of the law and the current ways in which air is included or overlooked, suggesting future developments for more integrated protection of this vital resource.

Chapter 6: *Thinking in a connected way*, highlights our journey towards systemic thinking, introducing the Ecosystem Approach and its implications. It details various case studies illustrating how some issues have been tackled in the past and how systemic insights could or should have led to alternative approaches, drawing out key lessons.

Chapter 7: *Rediscovering our place in the breathing space*, acknowledges a number of important principles and then applies them to a range of settings, including urban design, rural land use, value retention in material life cycles, novel economic approaches, and attitudes to risk and appropriate regulation, looking to the future and how we need to address the challenges we face.

Chapter 8: *Resolution for integrated management of the airspace*, argues that perhaps what we need is a set of consensual, science-based and high-level 'atmospheric principles' that recognise important aspects of the services provided by the air and the atmosphere. A set of such principles is proposed, emulating those that we already have for the integrated management of water and land resources, to guide the assimilation of this vital yet still commonly overlooked environmental medium in practical implementation of the Ecosystem Approach at scales from the biospheric to the international and intranational, and through to local and corporate settings.

The Annex reproduces the Millennium Ecosystem Assessment classification of ecosystem services and also key aspects of the Ecosystem Approach as a ready reference for readers who may be less familiar with these concepts and their sources.

The very notion that air and the atmosphere are an ecosystem will be novel to many, and the presentation of that ecosystem as the Earth's biggest and most neglected habitat will perhaps be surprising.

However, when one takes account of the breadth of services that the atmosphere provides to humanity, the scale and diversity of negative impacts that modern lifestyles exert upon it, and the lack of connections in the ways in which we seek to safeguard it, the importance of taking a more integrated approach becomes compelling. *Breathing Space: The natural and unnatural history of air* makes this case with solid proposals for further integrated development of society's interactions with this vital, yet massively undervalued, living heritage and natural resource.

1 | Air and the making of the atmosphere

The distinctions between the terms 'air' and 'atmosphere' are hardly airtight, even in scientific terms. Most definitions of 'air' relate to the gaseous content of the outer layer of the Earth. However, the term 'atmosphere' came into being in the seventeenth century as a compound of the Greek words *atmos* (vapour) and *sphaira* (sphere). So 'atmosphere' too relates to layers of gases that may surround a material body of sufficient mass, held in place by its gravity, though extending out in the case of planet Earth to include increasingly less dense and more energetic outer layers. For other planets that comprise mainly various gases, it is only the outer layer that constitutes their atmosphere. This book uses the term 'air' to refer to the physical 'stuff' we breathe and 'atmosphere' as the wider gaseous and energetic layers surrounding our home planet, although these definitions are necessarily porous.

The atmosphere is, by orders of magnitude, the largest habitat on our home planet. However, due to its fluidity, invisibility and lack of clear national ownership, it is also the most readily overlooked. The atmospheric system is not merely massive but also diffuse, comprising an immensely complex system of constantly flowing energy, matter and living things. Its frequent omission from our thoughts, and the fragmented way in which we have addressed isolated acute problems while disregarding impacts elsewhere in the air system, pose looming threats that we need urgently to address for our continued wellbeing. But first we need to understand it, and to develop a necessary degree of respect for the generally invisible yet endlessly protective and supportive bubble that we depend upon for our very life and livelihoods every single day.

What is air?

Answers to this deceptively simple question are complex and elusive. Dictionary definitions such as that of the *Oxford English Dictionary* – 'The invisible gaseous substance surrounding the earth, a mixture mainly of oxygen and nitrogen' – tell us a little more about the nature of air itself. In fact, at ground level, the approximate composition of air today is 78% nitrogen and 21% oxygen, the remaining 1% or so comprising a range of other gaseous substances. However, the properties of air change with altitude above sea level. With increasing altitude, air pressure reduces as the concentration of air molecules decreases, leading to what is called 'thin air', a major contributor to why many people suffer altitude sickness. In the upper atmosphere, where the air is thinnest of all, high in the stratosphere known as the 'ozone layer', there is a higher density of ozone molecules than anywhere else; these serve to block some of the sun's most energetic rays, deflecting potentially damaging radiation back out into space. Air does all of these things, and more, with us being barely aware of its existence.

> In the beginning God created the heaven and the earth. And the earth was without form, and void; and darkness was upon the face of the deep. And the Spirit of God moved upon the face of the waters.

These familiar words from the first chapter of the Authorised King James Version of the Bible, Genesis 1, articulate one view of the origins of our home planet and life upon it. Although only Creationists now tend to view this as in any way a literal representation of the origins of the Earth and its diverse life forms, the text is interesting in that it commits the same flaw as many subsequent scientific and popular texts: it omits to mention the air. However, as the Latin word *spiritus* means 'breath' or 'air', perhaps the intended meaning has been lost in translation into the dominant world view of 1611.

Even within the many pages of the United Nation's Millennium Ecosystem Assessment,[1] an authoritative study involving over 1,300 scientists in 95 countries in an assessment of the status and trends of all major global ecosystem types, air receives a relatively light touch. Chapter 13, dealing with 'Air quality and climate', comprises 36 pages within the overall massive 'Current state & trends assessment' report of the Millennium Ecosystem Assessment. A range of atmospheric 'services' are explicitly recognised in the chapter: these include warming; cooling; water recycling and regional rainfall patterns; atmospheric cleansing; pollution sources; and nutrient redistribution. The significant impact of both natural and managed ecosystems on climate and air quality is also recognised. But, in some ways, the air itself is largely a 'ghost at the feast', featuring by inference yet in reality always a fundamental medium linking all of the other 'major habitat types' assessed in the report. However, the wider benefits of the air to humanity are diverse and vital.

The only national-scale ecosystem assessment published at the time of writing, the UK's National Ecosystem Assessment,[2] ploughs a similar track by focusing on key habitats – 'mountains, moorlands and heath', 'semi-natural grassland', 'enclosed farmland', 'woodland', 'freshwaters', 'urban', 'coastal margins' and 'marine' – with no explicit consideration of air.

Unseen oceans of gas

Omission of the airspace as a major habitat type is surprising as it is not only important, it is also staggeringly big. Indeed, calculated up to an altitude of 100 kilometres – and including the atmosphere, the stratosphere, the troposphere and the mesosphere – the volume of air surrounding the planet is about 51 thousand trillion (51 trilliard) cubic kilometres, or 51,000 trillion trillion litres. This is about 38,000 times the volume of all of the world's oceans, which total about 1,347,000,000 cubic kilometres. The volume of the planet's air is so great that it eludes meaningful comparison in intuitive terms such as numbers of bathtubs, swimming pools or shopping malls.

Each of us draws in substantially in excess of 20,000 breaths per day, and we live our lives perpetually immersed in the bubble that surrounds the Earth. Yet, for something so immense and familiar to us, air remains incredibly easy to overlook.

That it is colourless, and also generally odourless and tasteless, blinds at least three of our five primary senses to it. We acclimatise to an ambient pressure of around 100 kilopascals (kPa) at sea level, equivalent to a pressure of nearly 14 pounds per square inch (psi). However, amazingly, we are unaware of this vast pressure except when we climb steeply, as our ears 'pop' or we experience altitude sickness, or when we immerse ourselves in the denser medium of water. And, although rapid air movement can be hugely destructive and fast movement through it can give rise to significant frictional drag, we can walk through still air without the slightest sense of its viscosity.

Of the five major senses, that leaves only hearing and the longer wavelength vibrations that we might sense as touch, both transmitted as physical waves. The rush of air is audible as it interacts with hard surfaces such as trees, buildings and ear lobes, and it is the physical structure of air that we depend upon to convey the pressure waves that we know as 'sound'. However, still air is not exactly the most raucous thing we ever experience.

To use a visual metaphor, the invisibility of air means that it is all too easily and frequently taken for granted, both in our daily experiences as biophysical entities and through our diverse industries and other activities.

Air, then, is a vital constituent of our home planet, or at least of the habitable part on the surface where all life lives, and is absolutely essential for virtually all life on Earth from plants to animals to microbes. And, of course, it is essential for us human beings who wade through it, generally unconsciously, every day.

How did all this air get here?

To vastly understate the reality, there is a lot of air about. So how did all this air get here?

The matter from which the solar system formed comprised mainly hot gases and dust circling a central core that condensed to create the sun. Most of the known planets of the solar system condensed from these same gaseous constituents; the best estimates suggest that our home planet Earth formed around 4.54 billion years ago. And that same matter, or at least virtually all of it, is still with us today.

Over geological timescales, heavier fractions began to settle out of this homogeneous amalgam through sedimentation, precipitation and other processes, progressively condensing into an increasingly solid internal structure as the Earth cooled. Due to gravitational forces, the most diffuse gaseous constituents formed an outer layer of the proto-Earth. They still do so today. The substantial size and density of planet Earth and its proportionally stronger gravitational pull, together with our location in the solar system, prevent the gases from becoming too warm and hence too energetic, and therefore enable our home planet to retain a durable atmosphere.

Clearing the air

Exposed as we are to the realities of accumulating waste and pollution stemming from the profligacy and lack of forethought that have shaped so many lifestyles and supporting technologies in the developed world, a common tendency is to think of the natural world as a pure canvas soiled by steadily rising contaminants. Taking a snapshot of the last two-and-a-bit centuries of history, this is undoubtedly true. However, a wholly different picture emerges if we take due account of the longer span of time over which the atmosphere formed.

As we have observed, the proto-planet constituted a largely homogeneous cloud of space dust, which progressively condensed into a solid core and outer layers of lower density. The gaseous composition of the primordial planetary atmosphere is thought to have been 98% carbon dioxide, 1.9% nitrogen and 0.1% argon. The cooling Earth became geologically active, with volcanoes spewing out vast quantities of lava, ash and gases, which then, as now, comprised mainly water vapour, carbon dioxide and compounds of sulphur, nitrogen and chlorine, with some molecules of methane and ammonia. This rich soup of atmospheric gases was conspicuously lacking in free oxygen, which, as we will see later, was a product of evolving life forms. The early atmosphere was therefore rich in what we refer to today as 'greenhouse gases' (the 'greenhouse effect' is described in greater detail later in this chapter), and was also 20 or 30 times as dense as it is today. Cumulatively, this resulted in heating the Earth's surface to temperatures as high as 85° to 110°C. Only as the atmosphere gradually cooled was water able to condense into clouds and then into liquid form as rains fell on bare rock, commencing the water cycle that in time would produce a covering of soil through the actions of both weathering and emerging life. Carbon dioxide would have been entrained by rain drops and also absorbed into early oceans, reducing the degree of warming through the 'greenhouse effect' and kicking off the exchanges of chemicals and energy that are the mainstay of biospheric cycles.

Today, we can broadly divide the solid Earth sphere into four layers. At the centre is a solid inner core, the hottest part of the planet comprising mainly iron and nickel at temperatures of up to 5,500°C. Surrounding this is the outer core, a liquid layer again largely made up of iron and nickel, and at temperatures similar to those of the inner core. The mantle lies around this; at approximately 2,900 kilometres thick, this is the widest section of the Earth. The mantle is formed of semi-molten rock called magma. While the rock forming the upper mantle is hard, the rock of the lower mantle is soft, much of it at the point of melting. The Earth's thin surface layer, between 0 and 60 kilometres of solid rock, is known as the crust. The terrestrial section of the Earth's crust is known as the continental crust, whereas regions covered by water are referred to as the oceanic crust. The oceans, though vast to our eyes, provide only a thin moist surface when compared with the sheer scale of the solid Earth, and they are dwarfed by the atmosphere above. The whole crust is made up of a network of tectonic plates that are in constant motion, and it is at the boundaries between these plates that most earthquake and volcanic activities ocur. All environmental media – earth, water and air – interconnect and play key roles in the great geochemical and energy cycles forming the 'biosphere': the domain in which life occurs and which life shapes.

It is within the solid matter of the planet that many substances we would now consider pollutants are locked away from the air and the cycles of matter and living processes above the Earth's crust. A great deal of this sequestration has occurred through sedimentation processes in which particles fall out of suspension in gas or air, sometimes flocculating first and then accumulating against a solid substratum. Over geological time, layers of sedimented particles consolidate into depositional landforms, progressively firming up into sedimentary rock. Mineralisation processes also lock various substances into solid forms that can become immobilised as rock. These mineralisation processes include the hydrothermal deposition of metals, many of them now economically important, into ores or 'lodes'. Hydrothermal circulation predominantly occurs close to heat sources in the crust of the Earth, including in proximity to volcanic activity; in such areas, granite intrudes in the deep crust due to orogenic processes, whereby land is uplifted at the intersections of tectonic plates. Hydrothermal deposits are also the result of metamorphic processes in which the properties of minerals change under the influence of heat, pressure and the introduction of chemically active fluids.

Together, these processes lock substances such as metals, phosphorus and other micronutrients into solid, buried strata of rock, effectively immobilising them and partitioning them off from free circulation in the biosphere. The net result is that, as the Earth has evolved, the biosphere has become progressively 'cleaner'. As concentrations of toxic and other biologically and climate-active substances have gradually declined, the biosphere has become increasingly more suitable for the emergence and evolution of complex life forms.

The structure of the atmosphere

The atmosphere's complex layers of magnetic and electric fields are subject to various scientific definitions and conventions. In most classifications, the different layers of the atmosphere are divided by major changes in temperature.

The inner layer of the atmosphere, from sea level to a mean approximate altitude of 12 kilometres (39,000 feet), varying with latitude, is known as the *troposphere*. This is the densest layer of the atmosphere, the weight of upper layers pushing down to create air pressure. Consequently, the troposphere contains 75% of all the gases in the atmosphere. The troposphere is also the layer that is habitable, taking into account the fact that there are no hard surfaces to live on at higher altitudes. It is also where weather patterns form, and hence the medium through which most natural water and other chemical and energy cycles occur. Nevertheless, there is considerable heterogeneity within the troposphere as, with increasing altitude, there is not only decreasing air pressure but also a decline in temperature of approximately 6.5°C for every kilometre elevation.

In addition to its gaseous composition, the Earth's atmosphere also contains a diversity of aerosols. Aerosols comprise colloids (fine particles normally invisible to the naked eye and suspended in another fluid medium) that include both solid and liquid droplets. These aerosols are of various types and concentrations, including both inorganic matter (such as fine dust, sea salt and water droplets) as well as organic materials (smoke, pollen, spores, viruses and bacteria, for example). Fine airborne particles that are either living organisms or are released by them are more commonly known as bioaerosols, shorthand for 'biological aerosols'. All aerosol particles are very small, ranging from less than 1 micrometre to 100 micrometres (a micrometre is one millionth of a metre), and so they remain suspended in the air for a long time, possibly indefinitely, reacting to air currents and potentially moving quickly and over long distances. Natural aerosols contribute to phenomena such as clouds and haze. The atmospheric load of aerosols therefore plays a significant role in water circulation, weather systems, the reflection of incident solar radiation due to the albedo (reflective quality) of clouds, the circulation of matter, and the distribution of and interactions between living organisms, climate and human health.

Above the troposphere, yet discrete from the stratosphere that lies above it, is the second true atmospheric layer: the *tropopause*. The temperature of the tropopause remains fairly constant, and it is here that very strong winds known as the 'jet stream' are found. The jet stream comprises narrow, fast-flowing currents of air moving from west to east, lying between 7 and 12 kilometres (23,000–39,000 feet) above sea level at the poles and 10 to 16 kilometres (33,000–52,000 feet) at subtropical latitudes. Jet streams, with speeds ranging from 90 kilometres per hour (56 miles per hour) to over 398 kilometres per hour (247 miles per hour), result from a combination of the planet's rotation and heating caused by incoming solar radiation. Jet streams have a major influence on weather and so are useful to weather forecasters. They are also exploited by aeroplanes as they can dramatically influence flight times if planes fly either with or against the flow of air, and they have been put to a range of other uses that we will touch on later in the book.

Above the tropopause is the lower part of the *stratosphere*, extending from an average altitude of 12 to 50 kilometres (39,000–164,000 feet). The temperature of the stratosphere remains fairly constant at around −60°C, and it is also here that the ozone layer is found. The ozone layer is formed as high-energy radiation splits oxygen molecules (O_2) into oxygen-free radicals (atomic O), some of which combine with oxygen molecules as ozone (O_3). Ozone therefore accumulates in this layer, absorbing ultraviolet radiation and some other ionising wavelengths of radiation emanating largely from the sun. The overall concentration of ozone is a result of the balance of energetic formation and destruction, but the net effect is that the ozone layer acts as a shield deflecting radiation from the Earth's surface. This absorption of energy causes the temperature to increase in the upper part of the ozone layer.

The *mesosphere* extends from 50 to 80 kilometres (164,000–262,000 feet), lying above the stratosphere. At this breakpoint, the temperature drops to about −100°C, the coldest region of the atmosphere. The mesosphere also serves a vital protective role: although the air is exceedingly thin (99% of the total mass of the atmosphere is below an altitude of 32 kilometres), it is here that meteoroids falling to Earth burn up.

The air of the *thermosphere*, the highest layer, above 80 kilometres (262,000 feet), is even thinner. However, its temperature is extremely high, often reaching in excess of 2,000°C, as ultraviolet radiation is converted to heat here. The thermosphere itself comprises two components: the first and inner one is the ionosphere, extending from 80 to 550 kilometres (262,000–1,804,000 feet) in altitude, and it is here that gas particles become electrically charged (ions) due to the absorption of ultraviolet and X-ray radiation from the sun. This phenomenon greatly supports radio communication, as radio waves are bounced off the ions and reflect back to the Earth's surface. The upper part of the thermosphere is known as the exosphere, which extends from about 550 kilometres for thousands of kilometres. Here, the air is very thin – the exosphere is virtually a vacuum. This is the layer in which satellites orbit.

Beyond the thermosphere – i.e. beyond the atmosphere above 1,000 kilometres (3,281,000 feet) – lies the *magnetosphere*. The magnetosphere, comprising highly ionised matter constituted mainly of protons and electrons, is in reality the outward extension of the Earth's magnetic field (also known as the geomagnetic field), which extends from where it is generated by the motion of molten iron alloys in the Earth's outer core (the geodynamo). The magnetosphere extends several tens of thousands of kilometres into space, where it interfaces with the solar wind (a stream of charged particles emanating from the sun) and cosmic rays from deeper space. The magnetic effects are concentrated into belts known as Van Allen radiation belts that trap particles and dangerous radiation carried in the solar wind. Without the magnetosphere, the solar wind would simply strip away the upper atmosphere, with devastating consequences. Collisions between the Van Allen belts and particles in the solar wind, particularly intense during solar flares, emit the characteristic shifting lights of the aurora borealis (the northern lights) and the aurora australis (the southern lights).

There is evidence of a consistent depletion of the Earth's magnetic field, commencing about 150 years ago; the overall reduction has been some 10% to 15%, but this trend has accelerated in recent times. The vital role of the magnetosphere in shielding the planet's surface and in helping a range of creatures navigate is leading some commentators to speculate that a reversal of polarity, a phenomenon that geological records reveal as occurring at a slow cycle over geological time, may be beginning. During such a polar reversal, the main field weakens and indeed almost vanishes, with a new field of opposite polarity generated by the circulating molten metal of the Earth's outer core then forming over thousands of years. The repercussions could potentially be catastrophic, certainly for many facets of modern life such as satellite communication – some of which is already compromised by the declining field strength – and possibly also for biological life. The last such flip in polarity occurred 780,000 years ago, at the time when *Homo erectus* was still learning how to make stone tools, so the impacts on life may not necessarily be apocalyptic, although potential implications may include depressed crop yields and a rising occurrence of cancers and cataracts in humans and fauna. The full implications are not only uncertain, but also beyond the scope of this book.

Life and the evolving atmosphere

So how did this great complexity within the atmosphere arise? To understand this, we have to look beyond simple physical processes: this complexity bears the fingerprint of life.

British scientist James Lovelock is well known as the prime proponent of what we now refer to as the Gaia hypothesis, along with the American evolutionary biologist Lynne Margulis. What is less well known is that Lovelock was, at that time, an atmospheric scientist, and his formulation of Gaia arose from NASA-sponsored research about the most effective means for detecting life on other planets. The presence of life on Earth, argued Lovelock, could be detected from afar by the instability of its atmosphere. For how else could such an inherently unstable condition – the abundance of highly reactive oxygen molecules as a prime indicator of instability – be maintained other than by the collective action of living things? The same principle should apply elsewhere in the universe, with the atmosphere of other worlds bearing an analogous 'fingerprint' of chemical instability where life exists, or may have existed.

The search for life elsewhere in the universe continues, but is another topic well beyond the scope of this book. However, the 'smoking gun' of the contribution of life to the structure and function of our own world is both fascinating and relevant. Lovelock's insights promoted the concept of a homeostatic Earth biosphere of closely co-evolved ecosystems and species, each contributing to and benefiting from the stability of the whole. This whole-system perspective is indeed central to thinking about the workings of the Earth system, and therefore for its sustainability.

Life itself made its evolutionary debut on Earth some 3.85 billion years ago. All organisms make their living by reacting with chemicals in their environment, inevitably changing the chemistry of the environment in which they exist. As a consequence, the arrival of living things instigated a new phase in Earth's evolution.

The origin of the oxygen-rich air with which we are familiar today began with what is known as the 'oxygen catastrophe', or the 'great oxidation event', about 2.7 billion years ago. Prior to this, the proportion of oxygen in the air was approximately only one-fiftieth of a percent (0.02%), or about one-thousandth of the contemporary level on Earth and around one-hundredth of the level present in the atmosphere of Mars. Like modern-day Mars, the atmosphere of early Earth primarily comprised carbon dioxide, which, as we have seen, made the Earth's surface a much hotter place.

Evidence from 2.7-billion-year-old shales in the Pilbara Craton of Australia is generally considered to show both eukaryotes (complex cells with nuclear membranes) and photosynthetic cyanobacteria (blue-green algae). Although there is some controversy about the exact timing of the appearance of oxygen-producing cyanobacteria, the 'great oxidation event' that caused the rise of atmospheric oxygen is certainly well documented in rocks laid down some 300 million years later. By 2.5 billion years ago, a group of microorganisms had evolved complex biochemicals and cellular mechanisms to harvest energy from visible light emanating from the sun, converting it into high-energy chemical bonds in molecules from which it could be released when needed. The arrival of photosynthesis – exploiting captured energy to fuse carbon dioxide and water into the simple sugar molecule, glucose, and emitting oxygen gas in the process – was nothing short of transformative for the evolving atmosphere and for all life forms within it. From an early atmosphere lacking in reactive free oxygen, this new solar-powered chemistry shifted the balance of atmospheric gases inexorably towards the oxygen-rich atmosphere that we know today.

With the advent of oxygenic photosynthesis by this new generation of microorganisms, the free carbon dioxide in the early atmosphere was progressively consumed, creating the 'waste product' of oxygen. The oxygen catastrophe is clearly demarcated in the geological record by a number of chemical markers, such as the introduction of large amounts of oxygenated iron (rust in the form of banded iron formations). Although we now regard free oxygen in the atmosphere as an absolute requirement for most forms of contemporary life, the relatively sudden transition to an oxygen-rich environment must have had a devastating impact on life forms that had evolved in an essentially oxygen-free world. Therefore, this event will inevitably have wiped out virtually all pre-existing life forms.

However, from this point onwards, more complex life forms were able to evolve in synergy with each other. In part, this was made possible by the greater access to nutrition and energy sources resulting from oxidative respiration. But also a diversity of habitats across the shallow waters and terrestrial surfaces of this world was opened up thanks to the contribution of atmospheric oxygen to the complexity of the atmosphere, significantly including the stratospheric ozone layer shielding the Earth's surface from ultraviolet and other ionising radiation from the sun and the wider universe.

The blossoming of life was accelerated by plants adapting to free themselves from the need to be immersed in water, taking root on land as part of an explosion in both the diversity and abundance of living things. This in turn enabled the evolution and spread of herbivores, grazing animals nourished by solar energy and the chemical matter captured from the air by plants, in turn feeding predators as well as an increasing menagerie of decomposers returning solid matter to the evolving soils and feeding gases back into the atmosphere.

As we have seen, many of the substances we might now consider to be pollutants, including various metals, phosphorus and other micronutrients, had been 'scrubbed' from the early planetary atmosphere by sedimentation, mineralisation and a range of other processes. These purification processes were then substantially accelerated by the intervention of life. Biomineralisation describes the process by which living organisms incorporate minerals into their tissues to harden or stiffen them. Biomineralisation is extremely widespread: over 60 different minerals have been identified in different organisms, ranging from silicates in algae, carbonates in various invertebrate groups, and calcium phosphates and carbonates in vertebrate animals.[3] Biomineralisation is known to have occurred for at least 750 million years.[4] Bacterial activity has been implicated in the creation of deposits of metals, including copper, iron and gold. Although these processes occur mainly in aquatic and terrestrial environments, chemical substances cycle through all environmental media. So these biologically enhanced processes contributed to the long-term, progressive overall purification of the entire biosphere, including the atmosphere.

The process of life changing the nature of the atmosphere, creating greater synergies, did not stop there – and, indeed, it has not stopped since. For example, during the Carboniferous period approximately 250 million years ago, terrestrial plants thrived, resulting in oxygen levels that were even higher than they are today. This permitted the existence of very large insects including *Meganeura*, dragonfly-like creatures with a two-foot wingspan that would not be able to survive in today's atmosphere due to its relative lack of oxygen. This was the era of great carbon deposition: organic matter and its high-energy bonds forged by energy captured from sunlight became sequestered from the early atmosphere and locked away into deposits – peat, coal, oil and gas – in the growing crust of the Earth as dead creatures sank in the deep oceans or were buried as part of ancient forests. The biosphere of planet Earth was becoming an ever more complex place, but also ever 'cleaner' (in today's terms) as substances that were toxic to life forms became progressively isolated from the biosphere.

So the modern atmosphere is a product of the collective action of living things. Each organism shapes the environment of which it is an inextricable part, and is in turn shaped by that environment. The co-evolution of all life forms as elements of tightly interdependent ecosystems vastly accelerated the throughput and efficiency of the processing of material resources through the biosphere, and made possible the evolution of ever more complex life forms, of which humanity is but one example.

The continuing sustainability of planet Earth rests upon the diversity, adaptability and efficiency of the ecosystems that it supports and that have shaped it. To seek to understand any element of nature outside the context of the ecosystems with which it evolved and of which it remains an interdependent element, from the whole biosphere to microhabitat scale, is to disregard its very essence, origins and future dependencies. To overlook human interdependence with the highly complex socio-ecological system of planet Earth that gave rise to us, including the air and atmospheric systems, is to create dangers.

Life could not occur without the protective and supportive capacities of the atmosphere. However, it is also true that the many interactions between life forms have been key agents of the evolution and continuing stability – or perhaps 'dynamic instability' is a more apposite term – of the atmosphere.

Physical forces of change

Barely resisting us as we move through it at walking pace, it is perhaps surprising to think of air as a potentially powerful force. Yet much of the planet is sculpted by wind and the particles that it carries. Aeolian processes are named after the Greek god Aeolus, the keeper of the winds. They describe the ability of winds to shape the surface of the Earth, and indeed of other planets. There are diverse forms of aeolian processes, including direct erosion by winds as well as the transportation and/or deposition of materials.

Even though its greater density means that water is a far more erosive force than wind, aeolian processes are particularly important in arid environments such as deserts. Indeed, the sand-blasting effect of particles borne on strong winds can give rise to striking and characteristic land forms in these regions. Furthermore, where vegetation growth is sparse, and consequently soils are unconsolidated, winds may be highly effective agents in moving sediments around. Common salt (sodium chloride) is blown up from the sea's surface and in fine spray as waves hit the shore, carried on the winds and potentially distributed over long distances, where it changes the chemistry and influences the ecology of landscapes. Wind is indeed a significant force in the erosion of the Earth's surface, as turbulent eddy action picks up particulates or else harder surfaces are progressively abraded, particularly where there is sand-blasting from windborne particles.

Sustained aeolian erosion can create what are known as deflation zones: pavements of rock fragments left behind after wind and water have removed finer overlying particles. These finer particles enter the air column and can be transported at distances ranging from the local to the intercontinental, or even, for the finest particles, they can remain there indefinitely. Locally, dust raised into the atmosphere may cause natural dust storms, particularly in desert regions where fine particles are most exposed. The sand dunes of desert systems are striking examples of land forms created by the deposition of mobile, wind-blown particles. Mass aeolian transport of fine particles may also be important for the redistribution of nutrients and other substances around the globe: for example, some minerals are recorded as having been transported from the Sahara to Amazonia,[5] whereas fine Saharan sand is regularly deposited over northern Europe.[6]

Meteorological marvels

Weather systems vectored by the troposphere too are among the fundamental processes shaping the Earth. Weathering processes break down rocks and soils into ever smaller fragments, and eventually to constituent substances. Rainfall absorbs carbon dioxide from the surrounding air, causing rainwater to be slightly acidic, thereby augmenting the water's capacity for erosion and dissolution. Released sediment particles and constituent chemical substances then enter into circulation in the biosphere.

However, to understand weather, which describes the instantaneous condition of these variables over shorter timescales, one first has to understand the climate system. The term 'climate' – derived from the ancient Greek *klima*, with meanings including 'slope' and 'region' – relates to long-term patterns and variations in a range of meteorological variables. These include atmospheric pressure, wind, temperature, humidity, precipitation and density of particulates within any given region, and these factors interact with local conditions to produce localised weather. As put succinctly by the US National Weather Service Forecast Office for Tucson, Arizona: 'Climate is what you expect, weather is what you get.'[7] Climate varies substantially with latitude, altitude and terrain, as well as with proximity to water bodies, which have a generally moderating effect as well as acting as sources of moisture.

The climate system is one of the most important aspects of how air and higher levels of the atmosphere create and maintain equable conditions for life on our home planet. Naturally, the climate system is interactive, driven by five principal internal planetary components. The first of these is the atmosphere, with other components being the hydrosphere, the cryosphere, the land surface and the biosphere.

The *hydrosphere* comprises all fresh and saline water surfaces as well as subterranean groundwater. The water cycle shapes soil moisture and the composition and circulation of the oceans, which, as they cover approximately 70% of the Earth's surface, play a major role in storing and transporting not only large amounts of energy but also dissolved carbon dioxide. As water is far denser than air, oceanic (so-called thermohaline) circulation is substantially slower than the atmospheric circulation, exerting an overall dampening effect that regulates the Earth's climate. Interactions with water vapour in the atmosphere are also clearly highly significant for climatic stability.

The *cryosphere* includes ice sheets, glaciers, snow fields, sea ice and permafrost. It plays a role in climate regulation as a thermal inertia, because it stores vast volumes of water and because it is a driver of deep oceanic currents, but most significantly through its high reflectivity, or albedo effect. A high albedo reflects back a greater proportion of the heat arriving as solar radiation.

The *land* surface, not in solely geological terms but also including vegetation and soils, exerts significant control over the absorption and return to the atmosphere of energy received from the sun. The return of infrared (long-wave) radiation from the land's surface has a substantial warming effect on the atmosphere. It also brings water into the atmosphere through evaporation from water bodies and moist surfaces and through evapotranspiration (evaporation by vegetation), as well as dust. These effects are modified by the topography and ecology of the land surface.

The *biosphere* includes both aquatic and terrestrial living systems, which have a major influence on the composition of the atmosphere. For example, photosynthetic processes absorb significant quantities of carbon dioxide and store them in living and dead tissues. These processes include those that sequester carbon into the Earth's crust, such as those observed during the Carboniferous period. Living systems also play a major role in the cycling of carbon and nitrogen, and hence in the emission of climate-active substances such as methane and nitrous oxide, as well as of aerosols and volatile organic compounds that have an important effect on atmospheric chemistry.

All of these planetary components interact through a diversity of physical, chemical and biological processes, thereby adding to the complexity of the climate system – this complexity includes both the climate's spatial and temporal variability and also its net stability as described by the Gaia hypothesis. The atmospheric and oceanic components are particularly strongly coupled through the exchange of parameters such as water vapour, heat and gases. These exchanges contribute to important climatic processes such as evaporation and condensation, cloud formation, precipitation, run-off and energy transfer through weather systems. One example that may be particularly significant for the transport of water vapour over long distances is 'atmospheric rivers';

these are plumes of concentrated moisture in the atmosphere that may be several thousand kilometres long yet only a few hundred kilometres wide.[8] A large atmospheric river may carry a greater flux of water than the Amazon, the planet's largest river, and atmospheric rivers account for over 90% of global north–south water vapour transport.[9] Atmospheric rivers are becoming increasingly implicated in localised serious flooding events.[10] Also, somewhere between 50% and 85% of the oxygen content in the air we breathe is produced by oceanic phytoplankton, tiny plants suspended in seawater.[11] The complex interactions between planetary components constitute a diversity of feedback systems. The climate system is also shaped by external forces, including, significantly, the sun, which is a dominant force impinging upon all planetary components.

The influence of the climate system on life on Earth can hardly be overstated, and, as we have seen in considering the 'oxygen catastrophe' and Gaia, this influence is reciprocal. There are various means to classify climate, but most commonly these relate to temperature and precipitation: for example, the hot and semi-arid environment of much of southern Africa compared with the temperate and moist climate of northern Europe. Climate also varies over time, including from season to season, from year to year, between decades, and over significantly longer timescales, as exemplified by the ice ages. Short-term variability is the norm. However, longer-term persistent trends are referred to as 'climate change', and these may be due to natural or human-driven forces. We will turn to human-induced climate change later in this book; however, first, we need to understand the natural processes driving climate change.

A natural greenhouse

The atmosphere comprises a structured yet fluid and rapidly changing body of gases and energy flows. The dominant gases, nitrogen and oxygen, have only limited interaction with incoming solar radiation and the infrared radiation that is re-emitted from the Earth's surface. However, various trace gases, known as 'greenhouse gases', both absorb and emit infrared radiation.

The term 'greenhouse effect' was first coined by Swedish scientist Svante Arrhenius, although the concept was first predicted in 1827 by French mathematician Joseph Fourier. Greenhouse gases are so named because, by absorbing infrared radiation emitted by the Earth and re-emitting it as heat, they tend to raise the temperature of the atmosphere near the Earth's surface. The result is a rise in the Earth's temperature above that which could be predicted from its distance from the sun alone, as we have seen when considering some implications of the 98% carbon dioxide concentration of the early terrestrial atmosphere.

This warming effect enables the existence of water in its liquid form on this planet, which itself supports the genesis and proliferation of living things. Nature's cycles maintain greenhouse gases in balance, contributing to the stability of the atmosphere. The main naturally occurring greenhouse gases in the atmosphere are carbon dioxide, methane and water vapour, although the effects of nitrous oxide and ground-level ozone are also significant. The role of ozone in the atmosphere is particularly complex: although it is a pollutant and climate-active gas in the lower layer of the atmosphere, the naturally occurring layer of high ozone concentration in the stratosphere has an invaluable function in absorbing and filtering out a harmful excess of solar ultraviolet radiation.

Although only some 0.001% of the global water resource occurs in the atmosphere,[12] it plays a significant role in regulating the planet's temperature. Water vapour is the strongest of the common greenhouse gases, and so its abundance and distribution in the atmosphere has a major impact on climate and the trapping and dissipation of energy within it. Consequently, clouds also play an important part in the natural greenhouse effect. This is not only because water vapour absorbs and emits infrared radiation, but, more significantly, because the brightness of most clouds creates a high albedo that reflects solar radiation with a net cooling effect on the climate system. This scale of this effect varies, depending on the height, type and optical properties of the clouds.

Forecasting the weather

So the air system and other elements of the atmosphere are the locus of a climatic system vital for the maintenance of conditions suitable for life, which in turn are influenced by life. They are also the medium that circulates energy, water and chemical elements such as nitrogen and carbon, and that carries spores, pollen and seeds as well as scents and pheromones.

Natural variations in the climate system occur over both long-term and short-term cycles. Examples of long-term cycles include the El Niño-Southern Oscillation (ENSO), which results from the interaction between atmosphere and ocean in the tropical Pacific, and the North Atlantic Oscillation (NAO), which has such a strong influence on the climate of Europe and parts of Asia and is driven by variations in barometric pressure near Iceland and near the Azores. There are even longer-term alternations of glacial and interglacial periods as a result of variations in the Earth's orbit, and the longer-term variability introduced by plate tectonics. A number of shorter-term feedback systems also occur, allied with the inherently chaotic (in the scientific sense of unpredictable) behaviour of such complex systems.

The term 'weather' describes our day-to-day experience of localised variability in wind, precipitation, cloud cover, sunshine, hurricanes, dust storms and a range of other rapidly forming and decaying parameters resulting from the chaotic climate system. The climate system itself integrates the effects of principal driving components on the troposphere, including trends in multiple layers of the atmosphere modified by proximity to water bodies and by land topography. Key driving forces of weather include local variances in air pressure as well as in temperature and moisture, influenced by a range of factors such as the angle of incidence of sunlight and jet stream patterns. To give a well-known example, the tilt of the Earth's axis and its influence on the incidence of solar radiation are major driving forces of seasonal weather patterns. The inherent complexity of the climatic system and its interdependence with a range of natural and anthropogenic factors mean that the weather systems that determine the local weather patterns that we experience on a day-to-day basis are often unpredictable. Extreme weather events such as heat waves, prolonged cold snaps, droughts and floods can occur as a result of natural variability, although the probability of their occurrence may potentially be exacerbated by human-induced factors.

The great natural cycles

The atmosphere, then, is a living, evolving and interactive system. The natural history of the atmosphere shapes, and is in turn shaped by, living and non-living components. The natural cycles that maintain the structure and function of the atmosphere are shaped in large part by the actions of life, while fluxes of energy, water and a range of chemicals in turn influence life. Life and the quality and structure of air and the atmosphere are indivisible.

The importance of the water cycle to all life, to meteorological cycles and to atmospheric shielding means that its role in the atmosphere as a key vector will be touched on many times throughout this book. Air also mediates aspects of silicon, phosphorus and other biomineral cycles, including through aeolian erosion and the transport and deposition of dust and a range of other substances and energy.

The interplay of oxygen and carbon dioxide in the great storehouse of the atmosphere is also part of a closely interlinked set of carbon, energy and oxygen cycles, interfacing with those of other substances and mediated by photosynthesis, respiration and decay. The nitrogen cycle deserves particular attention in this chapter because the largest environmental pool of the element is in the air, which today comprises 78% nitrogen. The nitrogen cycle is the process by which nitrogen is converted between chemical forms due to a number of linked biological and physico-chemical processes. Atmospheric nitrogen occurs mainly in a molecular form (N_2) that has limited availability for biological processes. Nitrogen is a key element in the structure of amino acids, proteins and the nucleic acids from which deoxyribonucleic acid (DNA) and ribonucleic acid (RNA) are formed, and therefore it often becomes a key micronutrient limiting natural productivity.

Some bacteria can fix gaseous nitrogen into more biologically available forms, and these bacteria may associate with other life forms, such as the root nodules of plants including those in the pea family, *Fabaceae*. For this reason, plants such as clover and lucerne are often planted as forage and ley crops to regenerate the nitrogen content of soils. Some fixation occurs in lightning strikes, but most is done by free-living or symbiotic bacteria. The biological fixation process plays a key role in converting gaseous nitrogen into biologically available forms such as organic nitrogen, ammonia and ammonium (NH_3 and NH_4^+), nitrite (NO_2^-), nitrate (NO_3^-), nitrous oxide (N_2O) and nitric oxide (NO). Other important transformation processes in the nitrogen cycle include nitrification, in which ammonia is oxidised into nitrate primarily by soil-living and other nitrifying bacteria; this usually involves a two-stage transformation by different organisms, with nitrite (NO_2^-), a toxin if allowed to accumulate, as an intermediate product. Denitrification is the microbially vectored reduction of nitrates back into largely inert nitrogen gas, completing the nitrogen cycle. There are obviously significant additional pathways and complexities to this cycle in different habitat types. However, this overview of the nitrogen cycle emphasises air as a key reservoir, and highlights the role of ecosystems in mediating and limiting the throughput of this essential life support cycle through aerial, terrestrial and aquatic media.

Land, water and air

Air and the atmosphere, then, are far from being distinct from each other, but rather act as a great continuum connected by and connecting the Earth's great cycles of matter and energy.

Take forests as just one example. Walking into woodland, even a small stand of trees in an urban setting, one is struck by a number of things. There is the rustle of leaves and the twittering of birds, the stillness of the air as the canopy of leaves baffles extraneous noise and quietens turbulent air currents. There is also the coolness, since trees breathe out not merely oxygen but also water wicked up from deep underground.

In fact, forests breathe water into the air in huge quantities. In the heart of Africa, as much as 90% of the rainfall of the extensive rainforest of the Congo basin is generated by the forest itself. The rainforests of the Amazon, comprising half of the planet's remaining tropical forests, not only are home to one in ten known species on Earth but also play a major role in regulating the regional and global climate. Part of this role arises from the fact that they contain 90 billion to 140 billion tonnes of carbon.[13] Furthermore, the Amazonian rainforest, though fed extensively by rainfall from external sources, including water vapour blown in from the Caribbean and the tropical ocean, also recycles up to a third of its moisture content, mitigating the effects of drought and also affecting adjacent ecosystems.[14] The role of forests in stabilising the climate and rainfall patterns at local and global scales cannot be overestimated. Trees, both individually and as forests, from the minute to the massive, connect earth, water and the air in many ways.

The oceans, too, play a major role in storing and recycling carbon, as do the world's wetlands, soils and other major habitat types.[15] The mass of stored carbon, heat, water, nutrients and other substances makes the integrity and functioning of these habitats important in regulating the atmosphere, as are biologically and functionally important ecotones (habitats at the interfaces between the principal environmental media).

The deep interdependence of all things

The atmosphere structures life, and life structures the atmosphere. The Earth system comprises deep interdependencies between all living and non-living things, including the media of air, land and water.

As we have seen, this is the basis of the Gaia hypothesis, the concept of a homeostatic Earth biosphere of closely co-evolved ecosystems and species, each contributing to and benefiting from the stability of the whole. Life is indeed essential for the sustainable workings of the Earth system, which include the maintenance of an otherwise inherently unstable atmosphere that shields, waters, provides yet limits nutrient substances, and circulates endlessly to maintain a viable 'bubble' within which life can prosper in the face of the intense cold and hostile radiation of space.

2 | Living in a bubble

We may think of ourselves as beasts that walk the surface of the Earth, but, in reality, we walk deeply within an atmospheric sea with which we co-evolved and from which we are indivisible. Even the 536 people from 38 countries who have gone into space, of whom only 24 have travelled beyond low Earth orbit,[1] have had to take with them a compressed source of home atmosphere or else some technical means of regenerating spent air. We are integral to the Gaian bubble into which we were born and which has shaped our bodies and psychologies, just as we, as cogs in the marvellously complex 'machine' of the biosphere, inevitably shape that biosphere and all the living and non-living things with which we share it.

This concept was grasped by Plato some 2,000 years ago in his dialogue *Phaedo*:

> for we are dwelling in a hollow of the earth, and fancy that we are on the surface; and the air we call the heaven, and in this we imagine the stars move. But this is also owing to our feebleness and sluggishness, which prevent our reaching the surface of the air: for if any man could arrive at the exterior limit, or take the wings of a bird and fly upward, like a fish who puts his head out and sees this world, he would see a world beyond; and, if the nature of man could sustain the sight, he would acknowledge that this was the place of the true heaven and the true light and the true stars.[2]

Our biological indivisibility

Assuming an average of between 12 and 17 breaths per minute at rest, and as many as 80 when we exert ourselves, we breathe the atmosphere in and out at the very least 20,000 times (in fact probably in excess of 30,000) each and every day. People over the age of 20 will have taken at least 100 million breaths.[3]

The biological infrastructure enabling this daily miracle is truly awesome. It is best appreciated by the simple experiment of trying to hold your breath and realising that, sooner rather than later, your body gives you no choice but to gasp in fresh air. Taken outside our accustomed breathing spaces, we may suffer altitude sickness due to the relative paucity of oxygen at higher elevations; this may even be fatal if we ignore the signs and symptoms. For most people, two or three minutes without air results in irreversible brain damage. Death ensues after four or five minutes' deprivation of this fundamental biological resource.

All vertebrate brains share a common underlying form based on three swellings of the neural tube, which are seen most clearly during early embryonic development. These three swellings develop into the forebrain, midbrain and hindbrain. These three portions remain about the same size in relation to each other in fishes and amphibians, but the forebrain becomes very much larger in higher vertebrates such as mammals, in which the overall size of the brain also increases. However, for all the sophistication enabled by the expansion of the cerebral cortex of the forebrain, the brainstem (the posterior part of the brain) retains subconscious control of vital bodily processes such as the regulation of cardiac and respiratory functions as well as maintaining consciousness, controlling eating and regulating the sleep cycle. Thus, breathing is controlled by the most ancient part of the brain, which evolved long before the arrival of consciousness. Whether awake, distracted, asleep or unconscious, breathing is of such fundamental importance that the unconscious brain puts its control beyond conscious meddling.

As our bodies are compelled to draw in breath, air surges deep down our throats right into the finest spaces of our two lungs. Human lungs are about the same size as a pair of footballs, yet their internal surface area of moist blood-rich tissue available for gas exchange is a massive 70 square metres (750 square feet), more or less the area of a tennis court. Air is drawn into and then squeezed out again from the lungs by the diaphragm, a smooth muscle under the lungs, and a network of intercostal muscles between the ribs. A deep breath can expand the lungs by 3 or 4 litres. Large particles are filtered from the inrushing air by fine hairs and mucous in the nostrils, which also moisten and warm gases passing through the nasal chamber and over the olfactory organ, where odours are sensed. Air then rushes downwards into the trachea, or windpipe, that branches into two bronchi, each of which supplies one of the lungs. The bronchi in turn divide into increasingly fine segmental bronchi and bronchioles, enabling the air to be carried deep within our chests to feed approximately 300 million fine grapelike sacs, known as alveoli, in each lung. The alveoli are surrounded by a rich network of fine blood capillaries and coated in surfactant substances that reduce surface tension, permitting air to exchange with around 25 billion red blood cells in the bloodstream. Haemoglobin pigments in these red blood cells bind with oxygen and carry it throughout the body.

Subconscious control of this process by the hindbrain is backed up by a wide range of other safety measures. The most externally obvious of these are the aversion that we develop to noxious or unknown smells, and also the largely subconscious coughing reaction triggered by irritation to the airways. Particles filtered by mucous and fine hairs are expelled by sneezing, another activity under largely unconscious control. Deeper within the body, a network of sensors include oxygen chemoreceptors in the aorta and carotid arteries, with additional carbon dioxide and acidity chemoreceptors monitoring the blood. These receptors feed inputs to the hindbrain, which then sends messages to the muscles of the diaphragm and ribs to increase the breathing rate in order to regulate the gaseous composition of the blood. Additional mechanical sensors guard the airways, including stretch receptors in the lungs that send signals that regulate the depth and duration of breathing. Other receptors monitor and control muscular activity, with feedback from emotions such as anxiety as well as from physical conditions such as yawning, sneezing or responses to pain. Overall, a comprehensive set of subconscious checks and balances from our very deepest subcellular level and in our nervous structure and brain programming maintains the passage of air into and through our bodies. That air is circulated by blood to nourish all our cells with oxygen and to carry away waste, particularly carbon dioxide and other breakdown products, maintaining our intimate connectedness with the air.

The great connector

Air is therefore also what connects us with all other life. Plants photosynthesise, regenerating the oxygen that we draw into our lungs and bloodstreams to nourish our cells. Reciprocally, the waste gases that escape our cellular metabolic processes via the bloodstream to be exhaled into the atmosphere are used and regenerated by the other organisms with which we share this common 'breathing space'.

In a thought experiment, the astronomer Harlow Shapley observed that, since around 1% of each breath that we take comprises the inert gas argon, we inhale and exhale 3×10^{19} atoms of the gas; these are then dispersed rapidly, so that they become smoothly distributed throughout all the free airspace of the Earth within a year.[4] A breath taken a year hence, wherever you are on the Earth's surface, will include at least 15 of the argon atoms of a breath you take today. The same is true, given the inert nature of argon, of the air exhaled by every great person, despot or nonentity that has ever lived, including both their first and their dying breaths. The very act of breathing binds us to all people and living things from the past and into the future.

As the genetic biologist and broadcaster David Suzuki put it, air is a 'universal glue' that joins all life. Every breath is therefore:

> an affirmation of our connection with all other living things, a renewal of our link with our ancestors and a contribution to generations yet to come. Our breath is a part of life's breath, the ocean of air that envelops Earth. Unique in the solar system, air is both the creator and the creation of life itself.[5]

Suzuki continues:

> As we imbibe this sacred element, we are physically linked to all of our present biological relatives, countless generations that have preceded us and those that will follow. Our fate is bound to that of the planet by the gaseous exhausts of fires, volcanoes and human-made machines and industry.

Air and the human senses

Sound is the term we allot to vibrations transmitted by the air at frequencies our ears have evolved to sense. Typically, this audible spectrum extends from about 20 Hertz to 20,000 Hertz (wavelengths of 17.2 metres to 17.2 millimetres) for humans. Other species have different, generally overlapping, audible ranges. For example, cats have a similar range at the low end of the scale but their acuity extends up to 64 kilohertz, or about 1.6 octaves above the range of a human.[6] There are many species of bat and as many variations in the ways they echolocate, transmitting and receiving echoes of higher-frequency sound waves through the air to create an image of surrounding objects, with some species hearing sounds at frequencies of up to 110 kilohertz.

At sea level, air of 20°C (68°F) transmits sound at approximately 343 metres per second (1,230 kilometres per hour, or 767 miles per hour). The physics of acoustics are complex, yet such is the evolutionary refinement of our senses that we pay barely any heed to them, nor indeed to the fact that it is the air that makes such communication possible. Sound, after all, cannot travel through a vacuum.

Clearly, humans can, and some do, function without speech and/or hearing. But the richness of communication and expression of emotion that is transmitted by sounds, and the awareness it gives us of the approach of traffic and weather patterns, and our enjoyment of music, accent and laughter enrich our lives. For other species, the sense of sound is used to detect danger, to navigate, to predate and to avoid predation, and for various forms of communication. Humans are far from being the only species with sophisticated organs for launching sound waves into the air, as many frogs, birds, marine and terrestrial mammals and insects possess specialised organs for this purpose. And, of course, virtually all of the Earth's natural phenomena – flows of water and air, fire, rain, the crash of surf and earthquakes – produce characteristic sounds helping us understand and navigate around our surroundings in the airspace.

Noise, too, can perturb our world. 'Noise' itself has an engineering meaning relating to the obscuring of a signal. But the auditory noise of a motorway or the scratch of fingernails down a blackboard can also create unwonted stresses, just as olfactory and visual noise (smog and light pollution are examples of each) can have the same effect.

The sense of smell is another service that the air provides. Olfaction is the means by which we detect chemicals blown to us on the winds. Specialised sensory cells in the nasal cavity of humans and other air-breathing vertebrates, and on the antennae of many air-breathing invertebrates, detect odorant molecules to which they are attuned, translating them into a sense of smell. Mammals have about a thousand genes that code for odour reception, a fact that illustrates the importance of olfaction.[7] Humans have far fewer active odour receptor genes than other primates and other mammals.[8] However, female humans have a stronger sense of smell than males, female olfaction becoming significantly stronger around the time of ovulation, which suggests that it plays some role in reproductive behaviours.[9]

A wide diversity of air-breathing species uses smell to detect predators and would-be prey and mates, often over great distances. Leeches and fleas, for example, sense potential host organisms largely through detecting carbon dioxide and other species-specific odours. The tendrils of plants are especially sensitive to airborne volatile organic compounds, through which they 'smell the air'. Parasitic plants, such as dodder (various green plants of the genus *Cuscuta*), use this sense to locate their preferred hosts, and to lock onto them.[10]

For many mammals and reptiles – although not humans, it is believed – there is also an 'accessory olfactory system' in addition to the main olfactory system, the accessory system used mainly to detect pheromones. Pheromones are essentially hormones secreted externally from the body of an organism to trigger a social response in members of the same species. These may include alarm pheromones, food trail pheromones, sex pheromones, and various other types that affect behaviour or physiology. Pheromones are deployed by organisms as basic as unicellular prokaryotes (cells lacking nuclei with clearly defined membranes) through to insects, vertebrates and plants. Air and water are the fluid media through which these intraspecific signals are transmitted.

Air is also a colourless medium excellent for the transmission of light, although this is another facet that we take for granted. Sight is one of the dominant senses used by humans and other organisms to navigate, to hunt and evade, to signal aggressive, sexual, social and other forms of intent, and to comprehend our surroundings. Reading, body language, television and cinema, semaphore and billboard advertising are some of the ways in which we transmit signals every day using the medium of air through which the light waves and changing colours travel, although we are barely aware of this.

Pigs are also said to 'see the wind', though this is more likely to be because they possess one of the most advanced senses of smell in the animal kingdom, which, in a way, means that they can 'see the wind'.

The divine wind

Given the significance of the air to all aspects of our wellbeing, it is hardly surprising that wind and other facets of the atmosphere feature as deities and iconic symbols in many cultures.

Gods of the wind appear in many cultures, with literally thousands of deities of the air and sky found all across the world. In Greek mythology, for example, Aeolus is the keeper of the winds. There are various legends and conflicting descriptions of Aeolus. One of the more interesting accounts features in Book 10 of Homer's *Odyssey*, in which Aeolus gives Odysseus a tightly closed bag full of the captured winds so that he could sail home easily to Ithaca on the gentle West Wind. Aeolus is also credited as the progenitor of the Aeolians, one of the four major ancient Greek tribes, who originated in Thessaly.

The Romans had a similar god of the winds, Venti. The Egyptian god of creation and of the wind was Amun, while Nut was the goddess of the sky and the god Shu took on attributes of the air as a calming and pacifying influence. The Chinese wind god was known as Feng Bo, whereas the Japanese have Fūjin, one of the oldest Shinto gods. And so it goes across the world, with Njörðr the god of the wind in Norse mythology, Vayu the Hindu god of wind, and Tāwhirimātea the clearly very busy Māori god of weather, thunder, lightning, wind, clouds and storms. In Christian mythology, Ariel is one of the Archangels, the term Ariel meaning 'lion or lioness of God', and is involved with healing and protecting nature. Oloron is the supreme deity, the god of the sky and of heaven, in sub-Saharan mythology, and Anshar and his son An are sky gods in Mesopotamian myths. And so the world is populated from the early Germanic sky god Teiwaz, the primeval Greek god of the sky Uranus, the sky father Rangi and Tane-rore as the personification of shimmering air in Māori mythology, and the sky god Virococha of Inca lore. There are many more examples of gods of the wind and air across all continents and cultures.

Though perhaps better known in the context of the suicide pilots who named themselves after it during the Second World War, the Japanese word *kamikaze* translates as 'divine wind'. The term came into more common usage around the time of the two attempted Mongol invasions of Japan in 1274 and 1281, major military efforts undertaken by Kublai Khan to conquer the islands of Japan after Korea had submitted to become a vassal of the Mongol empire. These two failures to invade Japan were significant in that they put a halt to the expansion of the Mongol empire, but also substantially contributed to nation-building in Japanese history. Not only did Japanese forces mobilise to successfully repel these invasions, but Mongol forces also lost as much as three-quarters of their troops and supplies at sea during the two attempted invasions as a result of major storms. These became known as the divine winds, or *kamikaze*.

Although, for many of us, wind chimes are mainly a form of garden ornament played by the passing of air, their origins are far more exotic. For example, in ancient Rome, chimes known as a tintinnabulum were hung in gardens and porticoes; the tinkling sound as the wind passed through them was believed to ward off evil spirits. A tintinnabulum, which could also comprise an assemblage of bells, often took the form of a phallus or a phallic figure, a symbol thought to guard against the evil eye and to bring good fortune and prosperity. We see the heritage of this superstition in bells used in churches, serving to call people to prayer and sometimes to sound the hours, but also symbolically to keep away evil spirits.

Wind chimes served similar roles in India during the second century AD, and later in China, where it became popular to hang wind bells that tinkled due to the action of the wind on their clappers at the corners of large pagodas. There is some contention as to whether these bells were there to discourage birds or evil spirits, but in all probability they had a dual role. Diverse forms of wind chime are still thought to bring good luck in various parts of Asia, including in Japan, where glass wind bells known as Fūrin have been produced since the Edo period (between 1603 and 1868 under the rule of the Tokugawa shogunate). Wind chimes are also used in feng shui (translating literally as 'wind-water'), the Chinese philosophical system of harmonising human existence with the surrounding environment, including the omnipresent invisible forces of chi that are believed to bind the universe, Earth and man together. Wind chimes remain in common usage today in the East to maximise the flow of chi.

Beliefs about the air as a divine or fundamental element are played out in the funerary arrangements of different cultures. A prime example is the Hindu *kriya karam* ceremony in which the body is cremated on a pyre of wood over flowing water as a means of returning it to the five elements. I have had the honour of attending *kriya karam* on numerous occasions in India. The spirit, the first element, has already left the body. The fire consumes the body, returning that which is of fire to the flames that engulf it. Ashes fall as the body is consumed, and water washes away that which is of water as the pyre is broken up at the end of the ceremony. That which is of air returns in the smoke.

This same principle is seen in the sky burial ceremony of Tibet, the Chinese province of Qinghai, Inner Mongolia and Mongolia. In this rite, the body becomes 'alms for the birds', a literal translation of the Tibetan *bya gtor*. The human corpse is incised then placed on a mountaintop, exposing it to the elements and animals. Predatory birds, as denizens of the air, are part of the return of the body, now an empty vessel following the transmigration of the spirit, to the elements that formed it. There is also a practical aspect as the ground here is too hard and rocky for the digging of graves, and the scarcity of timber means that cremation is reserved for high lamas and some other dignitaries.

These are just some of the myriad deities and sky gods; the list of cultures that ascribe numinous significance to the air, the atmosphere and the heavens could go on and on as we mine deeper into the world's countless polytheistic and animist religions. The key issue here is that recognition of the significance of air for all life is a pan-global phenomenon, encountered throughout human history.

A vital element

Straddling a sense of the divine and of the mundane is an underlying conception that the substance of the world is made of basic 'elements', an idea at least as old as recorded history and one that probably existed long before that. Before modern chemical understanding and the use of the term to describe pure chemical substances distinguished by a unique atomic number, the word 'element' had other meanings that spanned the biophysical and the metaphysical.

The classical elements – often consisting of earth, air, fire and water, but varying with culture – occur in a diversity of beliefs and philosophies, comprising the simplest essential parts of physical and living matter and imbuing it with its fundamental properties and powers. These beliefs are inspired by extrapolation from the observed phases of matter, from the solid to the liquid, gaseous and fire, and have their earliest recorded roots in Babylonian philosophy. These four basic elements occur almost ubiquitously in models of the elements, although a fifth element, the ether, occurs in ancient Greek thought, in Hinduism – where it signifies that which is beyond the material world – and in Buddhism (along with two other elements) as a form of sensory experience. Tibetan philosophy includes a fifth element of space.

Chinese philosophy describes different types of energy in constant interaction with each other, which contrasts with the Western concept of different kinds of matter, and so recognises a different set of elements: fire, earth, metal, water and wood. The Chinese system omits air explicitly, although metal is associated with *Qián* (the sky or heavens) and wood with *Xùn* (the wind) and *Zhèn* (arousing or thunder). Japanese philosophy also recognises five elements, with air representing things that move.

The natural analogy with the phases of matter can be readily traced in pre-Socratic Greek thought, when air constituted one of many *archai* in a conceptual system that sought to reduce all things to a single substance. This influenced much subsequent thought, particularly medieval European beliefs including alchemy, when describing the properties of different substances. However, oriental belief systems such as those encountered in China, India and Japan are now regarded as more figurative than biophysically based.

Unsurprisingly, given its ubiquitous and mysterious nature, air features as an essential element in virtually all of these philosophies. A range of physical properties are ascribed to air – as well as medical ones in the Hippocratic definition of the humours of the body, in which blood is associated with the properties of air. For Homer, 'thick air or mist', later to become 'air', was one of the four elements from which the world is made. In Buddhism, air is assigned the sensory expression of expansion or vibration. In the philosophy of the seven chakras of the body, shared by Hinduism and Buddhism, air and wind are linked with the Anahata (heart) chakra.

Of all the four classical elements of ancient Greece and its daughter cultures, air is often seen as a universal power or pure substance, and therefore also as much metaphysical as physical. Its fundamental importance for all life is reflected in words derived from it, including, for example, words such as *aspire*, *inspire*, *perspire* and *spirit*, all of which come from the Latin word *spiritus*.

A fascination with the essential 'element' of the air has thus been a driving force of human attempts to understand the natural and supernatural workings of the world.

The shaping of places and people

Just as the structure and living forms of the Earth have been shaped by its agency, air has had a far wider influence in shaping our culture than merely in the realms of the divine and in other belief systems. Air has had a major and continuing effect in moulding landscapes and in contributing to the character of the places we call home, ranging from windswept cliffs, coast and mountains, the sculpting of trees by prevailing winds and the influence of moisture, particulates and gases in air currents on the flora and fauna that grow there.

The climate system of the troposphere has exerted, and continues to exert, a profound influence on the distribution of human populations and the diversity of climatic zones, which present different opportunities for human habitation and use. For example, desert regions have a naturally lower 'carrying capacity' in terms of water provision and crop production compared with oceanic temperate climates. Also, the rain shadow effect determines that the two sides of an island, continent, massif or other topographical feature will differ in suitability for habitation and for different types of crop, stock and other uses. However, although climatic diversity is a key factor influencing the highly variable spatial distribution of the global human population, the situation is complicated by factors such as proximity to coasts and major rivers, and is now heavily influenced by human modifications to the physical system in addition to broader cultural and socio-economic factors.[11] Half of the Earth's potentially habitable land area contains less than 2% of the human population at densities of less than ten people per square kilometre, while 3% of the potentially habitable land is home to high population densities of more than 500 people per square kilometre. In practice, physiographic parameters, particularly coastal zones and navigable rivers (although these may also reflect lower elevations), influence the localisation of human population more than climatic parameters do. However, where seasonal precipitation patterns are stable and predictable – for example, monsoons and river systems boosted by seasonal snow melt – the climatic system is more strongly correlated with population.

Climatic changes, such as the desertification of the Middle East and the formation of land bridges during glacial periods, have been responsible for human migrations. Other examples of climatic variations, such as the Medieval Warm Period (from about AD 950 to 1250) and the Little Ice Age (AD 1550 to AD 1850), as well as more localised extreme weather events such as typhoons and associated storm surges, have influenced habitation patterns.

The art of air

We have seen how the music of wind chimes has meanings both sacred and aesthetic. But music itself can be broken down, at its most basic level, to oscillations transmitted through fluid media, particularly the air but also water.

The evocative title of Rachel Carson's seminal book *Silent Spring* drew upon an apocalyptic vision of springtime returning without the music of birdsong. Humans have developed a wealth of technologies relating to the transmission of sound through airspace, ranging from musical instruments and music reproduction systems to telephones, sonar systems to image the world around us, radios and other broadcast media. Music also socialises us through conversation, and in terms of our sharing of music through festivals, concerts and the kinship of common tastes. Our social venues – pubs, waiting rooms, lifts and hotel vestibules – commonly have piped music to create effects as diverse as putting us at our ease and ushering us through at a given pace.

Famously, the artist Vincent van Gogh represented the energy of wind graphically in paintings considered among the most expressive ever created. The English romantic painter John Constable (1776–1837) is regarded as the most accomplished painter of English skies and weather, which he employs as the chief conveyor of atmosphere in his landscape paintings. Similarly, the landscape painter J. M. W. Turner (1775–1851) is commonly known as 'the painter of light' due to his fascination with the 'phantasmagoria' of light passing through the atmosphere; his paintings also include clear reproductions of the effects of the pollution arising from the ongoing English Industrial Revolution.[12] Visual culture, or 'cultural climatology', as expressed not merely through the paintings of van Gogh, Constable and Turner but also in the works of other artists, such as Claude Monet and Olafur Eliasson, highlights the cultural symbolism of skies and the weather.[13]

The language of air

The word 'air' seems to have emerged in the English lexicon from the Old French word *air* (atmosphere, breeze, weather), itself descended from the Latin *aerem* (air, lower atmosphere, sky) and the Greek *aer* (to blow or breathe). The word 'spirit' comes from the Latin source *spiritus*, meaning 'breath', 'air' and concepts such as 'inspiration' (or breathing in). The word 'expiration' derives from the same source, signifying our last breath, or the passage of the spirit from dense flesh into realms unknown.

Air also permeates our language. We speak of 'something in the air', a good or bad 'atmosphere', 'airing' an opinion or grudge, and so forth. In one way or another, all of these relate to communication through the notional ether or of moods hanging around us. To 'air' a grievance or an opinion is to express something publicly, and the 'air' of a person is the manner or appearance that emanates from them. And, of course, an 'aria', derived from the same source, is a melody or tune conveying a general mood in an individual.

In terms of creativity, we speak of 'blue skies thinking', signifying a breathing space in which new ideas can emerge. Conversely, classic songs such as 'Stormy Monday' and 'Stormy Weather' use weather systems as a metaphor for tempests raging in the mind or for strong emotions. Other songs, such as 'The Air That I Breathe', relate to freedom, as do many alluding to being free as a bird.

Air and technology

Technologies to harness energy from the air have made significant contributions to human progress over past centuries. Windmills, for example, have long been deployed to convert the energy of the wind into rotational energy by means of a set of sails. Originally deployed for the milling of grain, windmills have been used in much of the world for the pumping of water, including accessing groundwater and draining low-lying land, as well as for a variety of industrial purposes. Windmills date back a long way into human history, with the Greek engineer Heron of Alexandria providing the earliest known instance of using a wind-driven wheel to power a machine in the first century AD.[14]

Another early example of a wind-driven wheel is the prayer wheel used in ancient Tibet and China from the fourth century. Practical windmills with sails rotating in a horizontal plane were in use in Persia from the ninth century; these had sails of reed matting or cloth and were used to grind grain or draw up water. Their use spread through the Middle East, Central Asia, China and India, and into Europe during the eighteenth and nineteenth centuries.

The now more familiar horizontal axis or vertical windmill is believed to have originated in the twelfth century in northern France, eastern England and Flanders, predominantly for the milling of grain. A variety of continually evolving designs – including, for example, post mills, hollow-post mills, tower mills and smock mills – increased the versatility and application of windmills. Millers were initially able to adjust the amount of sailcloth spread over the lattice, but mechanisms for the automatic adjustment of sails using a series of shutters were developed in Britain in the late eighteenth and nineteenth centuries, leading to sails that adapted to the wind speed without the need for the miller to intervene. In addition to their continuing value for milling and the pumping of water, windmills found applications as sawmills, paper mills and threshing mills and to power many other industrial processes, including the processing of oil seeds, wool, paints and stone products.[15] The total number of wind-powered mills in Europe is estimated to have peaked at around 200,000,[16] declining from this figure with the onset of the Industrial Revolution as they were replaced by steam power and then the internal combustion engine.

Today, wind pumps are used extensively on farms and ranches in the central plains and the south-west of the United States, as well as in southern Africa and Australia, as a technology requiring little maintenance and deployed as stand-alone units that need no energy inputs beyond that provided by the wind. They serve in these often isolated settings as efficient means of pumping water and driving powered feed and agricultural machinery.

The modern descendants of the windmill are wind turbines, windmill-like structures specifically developed to generate electricity. They are widespread on farms in the United States, where centralised electricity distribution systems may not be available. More strategically, they are beginning to address emerging sustainability challenges elsewhere across the world, contributing to zero-carbon renewable energy and addressing the rising concerns over energy security, global warming and eventual fossil fuel depletion that have led to an expansion of interest in all forms of renewable energy generation. The 121,474 megawatts of installed wind energy capacity in Europe as of the end of 2013 represented 38% of the global wind generation capacity of 318,137 megawatts.[17]

At a higher altitude, there is also considerable research interest, if as yet no breakthroughs that are even remotely close to being marketable, in harvesting energy from the high-speed wind currents of the tropospheric jet stream.

Air also plays a major role in the combustion process. Perhaps we overlook how vital it is because the oxygen content of air is rarely if ever limiting. But it is air that oxidises the food we ingest, liberating energy to keep us going and growing, and it is also air that enables our bodies to burn in cremation ceremonies. It is air that combusts wood and, as the transition to more energy-dense fuels ushered in the industrial age, does the same for fossil fuels. Driving air at greater speed into the combustion process is the basis of blast furnaces and other advanced combustion techniques, smelting iron and other metals, which in turn is the basis for even more industrial progress. And, of course, the modern world has been profoundly shaped by the agency of the internal combustion engine, igniting air and liquid fuel to transport society, literally, into new opportunities for resource exploitation, trade and recreation.

Other uses of air include cleansing, for example in familiar household applications such as vacuum cleaners driven by fan or cyclone technologies. We also depend on flows of air for drying, be that the humble washing line or newfangled inventions such as Dyson's 'Airblade' technology, which uses focused jets of air to wipe water from skin surfaces.

It is hard to think of a contemporary factory without a compressor and pipe system servicing a wide range of functions. These may include using energy in compressed air to drive machines, to spray paint or other substances, to blow away dust and other dirt, for drying, or to propel conveyor belts. Air also serves as a means to propel other additives, be that in pressurised aerosol cans or in specialist applications such as the sterilisation of bottles with sulphur dioxide before they are filled with wine, jam or other food products.

A small proportion of the atmosphere is combustible – for example, methane released as a waste gas from processes breaking down organic matter in anaerobic environments such as wetlands. Today, methane concentrations are low, but about 3.5 billion years ago there was 1,000 times as much methane in the atmosphere, mainly due to volcanic activity, although the appearance of ancient bacteria added to the methane concentration by converting hydrogen and carbon dioxide into methane and water. With the evolution of photosynthesis and the increase of oxygen in the atmosphere, methane levels declined. Today, atmospheric methane is of interest primarily because of its impact as a greenhouse gas, trapping 25 times more heat per unit of mass than carbon dioxide.

Methane is also a key constituent of the mix of hydrocarbon gases that comprise natural gas, now widely used as an energy source for heating, cooking, electricity generation and some vehicles, as well as a chemical feedstock in the manufacture of some plastics and other commercially important organic chemicals. Natural gas is mined not from the atmosphere but from gas pockets and other hydrocarbon reservoirs in deep underground rock formations where it is created over time by both natural biogenic and thermogenic processes. Natural gas represents an efficiently transportable form of stored energy.

Modern methods such as fracking – artificial hydraulic fracturing of geological strata to release shale gas – are proving highly contentious due to their potential contribution to groundwater perturbation, earth tremors and unknown impacts on wildlife, all of which are perceived as being overlooked in the rush for commercial exploitation. The current commercial and political appetite for fracking and for other novel gas extraction techniques is perhaps most baffling given stated commitments across the world to progressively reduce dependence on fossil fuels en route to a carbon neutral future.

Perhaps the most pervasive of all uses of the atmosphere to support industrial and lifestyle purposes is the 'mining' of its capacities as a sink in which to disperse exhaust gases from all manner of transportation, combustion and other processes, without which modern lifestyles would simply cease. Its role in the dispersal of waste gases is therefore a fundamental if underappreciated and undervalued resource for industry, transportation, domestic heating and a huge range of other human activities. Barely considered as such, it therefore constitutes a major 'externality' – a value not factored in to economic considerations, including governance decisions, except where excessive local concentrations or emissions of certain hazardous, ozone-depleting or climate-active gases reach regulatory trigger levels. This is an area we will return to later in this book.

Air, exploration, trade and travel

Exploitation of the energy in the atmosphere has also been a significant propellant of our aspirations to transport ourselves and our goods. The origins of the sailing ship are lost in the mists of time, but all sail-powered vessels share the attributes of a hull, rigging and at least one mast to hold up sails that intercept the wind to power the ship. There are many types of sailing ships – mostly distinguished by differences in their rigging, hull, keel, or number and configuration of masts – and they have been a feature of coastal nations around the globe for all of recorded history. But try sailing without the wind to see how dependent this potentially sustainable method of transport is upon the air. Indeed, with growing concerns about climate change, fossil fuel depletion and resource insecurity due to political instability along supply chains, there is a resurgence of interest in sail technology as a more sustainable alternative or add-on.[18]

Aviation, too, is utterly dependent upon the air. Obviously, the internal combustion engine comes into play here to drive propellers that gain purchase on the fluid medium of air for forward propulsion. Jet engines also use air, in this case as an oxidation source for the controlled explosion of fuel to drive forward the aeroplane. But air plays a more profound part in suspending our flying vehicles, be they fixed-wing planes or helicopters. The phenomenon of lift makes use of the viscosity of air, generating aerodynamic force perpendicular to the flow of air passing over a wing or blade. Propellers use the same principle of aerodynamic lift to drive aircraft forward, and wings (spoilers) on racing cars act in reverse to generate downforce that presses the vehicle into the track to generate greater traction. The wing, propeller or aerofoil has a streamlined shape, convex on one surface and sometimes concave on the reverse side, generating lift as it passes through the air. The physics of lift are complex and indeed contested, and certainly beyond the remit of this book, but without air there is no lift. Air, then, provides us with the services, invaluable to the modern world, of suspending aeroplanes in the sky, much as hot air balloons stay aloft due to lift caused by different densities of hot gases inside and cooler gases outside the balloon.

Contrasting with the force of lift is that of drag. Drag occurs from the resistance of surfaces perpendicular to the direction of travel. Often, we consider drag as a hindrance to the forward motion of a car or aircraft. However, it is also a positive force dampening movement, making a major contribution to slowing down vehicles as well as enabling trees and other vegetation to absorb storm energy that might, if unabated, cause greater damage to buildings and associated built infrastructure.

In contrast to the use of air to generate lift is its channelling to counter gravity in the case of the hovercraft. Hovercraft are also known as air-cushion vehicles, and this describes how they use blowers to produce a large volume of air above atmospheric pressure below the hull or skirt. This enables the craft to float and travel over land, water, mud or ice with little friction. As a result, driving a hovercraft has more in common with flying an aeroplane than piloting a traditional boat. A British invention from between the 1950s and the 1960s, hovercraft are now used throughout the world as specialised transport vehicles. Examples have included large versions that transported hundreds of people and vehicles across the English Channel, others with military applications such as transporting tanks, soldiers and large items of equipment in hostile environments and over difficult terrain, and smaller vehicles that are invaluable for disaster relief, coastguard and survey purposes.

Another unsung benefit of air to transport can be found in the everyday pneumatic tyre. The pneumatic, or air-filled, tyre comprises an airtight outer layer, generally of metal-reinforced rubber, that is filled with air pressurised to greater than atmospheric pressure. This allows the tyre to remain inflated even with the weight of the vehicle resting on it, the compressibility and energy dissipation of the air in the tyre offering cushioning as the outer layer of the tyre hits bumps or dips in the road. This makes for a dramatically smoother ride.

Mining the air

While we may not think of it in these terms, the production of oxygen, nitrogen, the inert gas argon and other gases used for diverse commercial, medical, industrial, agricultural and other uses occurs by separating them from the air. There are no regulations on how much of these gases can be extracted, nor is any payment made for their extraction, perhaps due to the fluidity, lack of ownership and sheer volume of the medium of air.

The mining from the air of nitrogen, of all the gases, has become the most massive in pure scale. It is therefore worth paying nitrogen production a little more attention, and we will also consider some of the consequences of such massive fixation of the gas from the air later in this book. Nitrogen, let us recall, constitutes approximately 78% of air at ground level, yet, in its stable molecular form, it is unavailable to most living organisms. Although nitrogen is an essential nutrient for all organisms, as a key constituent of vital biochemicals such as proteins and nucleic acids, biologically available forms of nitrogen are relatively scarce in most environments. Nitrogen is therefore a crucial micronutrient, essential yet limiting productivity, and for this reason it is an important element of fertilisers added to land to enhance crop production. The importance of nitrogen to agriculture began to be appreciated in the late nineteenth century, after the reaction of atmospheric nitrogen with charcoal at high temperatures in the presence of alkali materials was first observed by the French chemist Desfosses in 1828. Nitrogen is also used extensively in the manufacture of explosives and a range of other products.

Much early production of fixed nitrogen used the Birkeland–Eyde process, developed in 1903 by the Norwegian industrialist and scientist Kristian Birkeland along with his business partner Sam Eyde. This process fixes atmospheric nitrogen as nitric acid, from which nitrate is then produced. However, due to its energy inefficiency, the Birkeland–Eyde process was gradually replaced by alternative processes. These included the Haber process in 1909, producing ammonia from methane gas and molecular nitrogen, and the Ostwald process in 1902 for the conversion of ammonia into nitric acid. The Haber process operates at high pressures of around 200 atmospheres and high temperatures of at least 400°C, producing ammonia from natural gas as a hydrogen source and air as a nitrogen source.[19]

The mining of nitrogenous fertilisers from the air is of great importance, both commercially and in supporting human needs. Today, about 30% of the total fixed nitrogen on Earth is manufactured in ammonia chemical plants,[20] dwarfing natural processes. The Haber process is now the largest source of nitrogen fixed into the Earth's ecosystem, and it is estimated that almost half the people on Earth are currently fed as a result of synthetic nitrogen fertiliser use.[21] The production of reactive nitrogen has doubled globally over the past century, and has tripled in Europe.[22] We will explore the 'downstream' effects of this later in this book.

Air and warfare

Almost nowhere has humanity shown greater ingenuity than in its attempts to kill each other, and so air has been used as a medium for warfare throughout the centuries. Most obviously, the sailing ships of the pre-industrial age, sails filled by oceanic wind, were a technological advance that greatly helped 'Britannia rule the waves'.

Poisonous gases released into the air as a weapon of war were a macabre innovation of the First World War, killing often indiscriminately and prone to causing disastrous self-inflicted harm when wind currents changed, carrying toxic atmospheric loads back to allied lines. Phosgene (a simple molecule comprising two chlorine atoms and one of oxygen double-bonded to a central carbon atom), chlorine in its molecular form and mustard gas (also known as sulphur mustards, comprising a range of organic substances with a central sulphur atom and terminal chlorine atoms) were all deployed from the First World War onwards by means of artillery shells, aerial bombs, rockets or spraying from warplanes. To this toxic arsenal was added a range of more sophisticated nerve agents. Most of these indiscriminate killers were outlawed under the 1993 Chemical Weapons Convention. However, this has not stopped their use by pariah regimes, such as in Saddam Hussein's Iraq and during the Syrian civil war of 2013, in which these weapons were deployed by ruling regimes against their own civilians. To this inglorious array of human creations has been added biological warfare, also known as germ warfare, in which biological toxins or infectious agents such as bacteria, viruses and fungi are dispersed through the air with the intent to kill or incapacitate humans, animals or plants as an act of war. Again, due to the indiscriminate nature of this type of weapon, some are covered by the Biological Weapons Convention while others fall under the Chemical Weapons Convention. Notwithstanding these conventions, there is growing concern about the potential for the clandestine poisoning of the air with these agents as an act of bioterrorism.

One of the more curious examples of the atmosphere being used for warfare was enacted by Japan, which exploited the recently discovered jet stream as a means to send fire balloons over the Pacific Ocean to the west coast of Canada and the United States towards the end of the Second World War. Fire balloons (*fūsen bakudan* or 'balloon bombs', also known as Fu-Go) were cheap if relatively ineffective weapons. The balloons were about 10 metres (33 feet) in diameter when fully inflated, holding about 540 cubic metres (19,000 cubic feet) of hydrogen. They could carry loads varying from a 12 kilogram (26 pound) incendiary device or a 15 kilogram (33 pound) antipersonnel bomb to four 5 kilogram (11 pound) incendiary devices. Between November 1944 and April 1945, Japan launched over 9,300 fire balloons from sites on the east coast of the main Japanese island of Honshū. Timed fuses, set according to estimates of when the balloons would arrive over American soil, triggered ignition of gunpowder to blow up the balloons and release the payloads. Around 300 fire balloons were found or observed in North America and they were responsible for one of the few attacks on North America during the Second World War. However, they caused only six deaths and a small amount of damage: a pregnant 26-year-old woman and five children between the ages of 11 and 14 were killed when they discovered a balloon bomb that had landed in the forest of Gearhart Mountain in southern Oregon on 5 May 1945. Nevertheless, they created disquiet for the simple reason that their arrival was mysterious. As nobody believed the balloons could have come directly from Japan, it was thought that they must have been launched by landing parties from submarines on American beaches, with even wilder conjectures that they could have been launched from German prisoner-of-war camps in the US or from Japanese-American internment centres.

And more factors besides . . .

It is clear from this brief review of the many ways in which air and the atmosphere have shaped our civilisations and progress that we have taken many of their benefits for granted. Some, such as their role in enabling us to release energy from organic matter for heating, disposal, power and transport, we may overlook as the air system is rarely limiting. Most others we overlook because air is invisible.

This is, let us remind ourselves, merely an overview of some of the many benefits that air provides to society. To document all such benefits would be tedious and would also exceed the illustrative purpose of this chapter. But one thing is for sure: our dependence on air is as total as it is almost totally overlooked.

3 | What does air do for us?

As if providing for our basic biophysical survival and health were not enough, air quite clearly does many other things for people. Many of the benefits that derive from air have been taken almost completely for granted throughout history. However, current rising population levels and the progressive degradation of ecosystems and the resources they provide, a conflict from which so many of today's sustainability concerns arise, are focusing minds on the need to better account both for impacts on ecosystems and for the many benefits they provide for us.

Frameworks for systemic understanding

Our consideration of the natural world and the benefits that flow from it has tended to be fragmented in the past. So too has been our response to addressing 'problems' such as local air pollution and damage to the climate or atmospheric shielding systems; we will return to this theme frequently throughout the remainder of the book. However, it is clear from the last two chapters that air and the atmosphere are not only internally complex but also deeply interconnected with all other environmental media, living and non-living elements of the biosphere, and human interests and activities. Failing to think systemically clearly robs us of important insights about the world in which we live, the benefits we obtain from it, our potential wider impacts upon it, and its capacity to support our needs into the future.

The idea of 'ecosystem services' arose in the late 1980s as a conceptual pedagogic and resource management framework with which to understand the diversity of benefits that the natural world provides to humanity. The concept of ecosystem services is defined by the Millennium Ecosystem Assessment as 'the benefits people obtain from ecosystems'.[1] In other words, ecosystem services classify what nature does for us. The ecosystem services framework also outlines the systemic nature of these benefits, and illustrates how exploitation of a particular service by one sector of society, or the perturbation of the ecosystems that produce that service, has wider ramifications for all other benefits and their diverse beneficiaries.

A wide range of ecosystem service classification schemes was developed from the late 1980s, generally addressing discrete habitat types within specific bioregions of the world, such as tropical wetlands, coral reefs, rainforests or temperate rangelands. The Millennium Ecosystem Assessment, a major United Nations study involving more than 1,300 scientists from 95 countries exploring the status and trends of major global habitat types and their implications for continuing human well-being, drew upon this variety of pre-existing categorisations. The result was a harmonised international classification scheme of ecosystem services that enables comparison between habitat types and across bioregions. This Millennium Ecosystem Assessment classification of ecosystem services provides a generically applicable assessment of the breadth of benefits that ecosystems provide to people, taking account of cultural diversity and associated values (both economic and non-economic). This classification of ecosystem services, reproduced in the Annex, recognises four principal categories of services:

- provisioning services, which comprise tangible, extractable and often tradable assets derived from ecosystems, including food, fibre, natural medicines, fresh water and energy;
- regulatory services, which refer to natural processes that regulate factors such as air quality, climate and microclimate, water purification, storm and natural hazard protection, diseases and pests;
- cultural services, which provide less tangible benefits, such as aesthetics and regional character, educational, tourism and recreational opportunities, and artistic inspiration; and
- supporting services, which comprise a range of processes that maintain the ecosystems' integrity, functioning and capacity to supply other services, such as soil formation, habitats for wildlife, nutrient cycling and primary production.

Ecosystem services are intentionally anthropocentric and classify outputs from ecosystems that are relevant to all human interests, ranging from basic biophysical health and wellbeing to the security of economic resources, broader 'quality of life' factors and the overall resilience of the socio-ecological system. It is this breadth – not only of benefits but also of value systems – that makes them inclusive of a wide range of human interests and of the many dimensions of our dependence on the ecosystems that make life possible, profitable and fulfilling.

Importantly, the harmonised international Millennium Ecosystem Assessment framework also reflects systemic interactions between elements of the environment, the services that it provides, and the ways in which service exploitation by a subset of people can affect the benefits enjoyed by others. For example, modern intensive food production systems (farming of land, capture fishing, aquaculture and so on) have substantially boosted production within a narrow section of provisioning services (particularly food and fibre) but at substantial, if unintended and largely underappreciated, costs. These costs include the release of carbon sequestered in the soil into the air, along with nitrous oxide, other aerial contaminants and water, as well as the exacerbation of aeolian erosion.[2] To illustrate the extent of unintended consequences from land use, an analysis addressed in the Millennium Ecosystem Assessment 'general synthesis'[3] concluded that the 'external cost of agriculture in the UK in 1996 (comprising damage to water, soil and biodiversity) amounted to some $2.6 billion, or 9% of yearly gross farm receipts'. The net costs of a narrow focus on production of any one service may therefore be substantial if the wider negative consequences for other services and societal constituencies, including future generations, are overlooked.

The Ecosystem Approach, promoted by the Convention on Biological Diversity (CBD) as 'a strategy for the integrated management of land, water and living resources that promotes conservation and sustainable use in an equitable way',[4] sets geographical and socio-economic contexts within which to consider the management of ecosystems and their services. Importantly, it recognises that humans and our economic and other activities are central and interdependent components of ecosystems. The CBD identifies 12 'complementary and interlinked' principles for the application of the Ecosystem Approach; these are elaborated in detail in the Annex, along with the background to the approach and its points of operational guidance. In Chapter 6: *Thinking in a connected way*, we turn to the relevance of the Ecosystem Approach for managing our interactions with the air and the atmospheric system. However, for the purposes of this chapter, the ecosystem services framework provides a conceptual basis to better understand the diversity of benefits that these systems provide to humanity.

Systemic assessment of the value of air

Despite the multiplicity and strength of interconnections between atmospheric processes, layers and services, there has formerly been little consideration of the sum total of services that they provide and hence no systemic analysis of the value of the air and the atmosphere to humanity. This is of concern, since the air system is evidently vital for so many aspects of human interest and endeavour.

Chapter 13 of the Millennium Ecosystem Assessment,[5] though primarily addressing air quality and climate change, as indicated by its title 'Air quality and climate', also lists a few additional ecosystem services provided by the air system (warming, cooling, water recycling and regional rainfall patterns, atmospheric cleansing, pollution sources and nutrient redistribution). Among the earliest systemic explorations of the ecosystem services provided by air and the atmosphere are studies by John Thornes and colleagues[6] – these identified 12 'environmental services' assigned to different layers of the atmosphere – and research by myself and by collaborators[7] that considered multiple ecosystem services grouped by the principal categories defined by the Millennium Ecosystem Assessment.

Building from these bases, but also taking account of the wider considerations covered earlier in this book, Table 3.1 (pages 50–1) lists a more extensive set of ecosystem services oriented by both atmospheric layer and category. In this table, services assigned by Thornes et al. are also aligned with service category, and services recognised by Everard et al. are associated with the most appropriate atmospheric layer (acknowledging some overlap between these 'environmental services' and ecosystem services), with additional services mentioned previously in this book included in the table where most relevant.

Lessons learned from this analysis

Although reductive analysis is helpful to identify key features of a system, we have to recall that systems are inherently interactive internally and that the interaction of constituent elements creates the 'emergent properties' that define the system. (Examples of the emergent properties of systems include: living processes, due to the interactions of cellular organelles; the catalytic properties of enzymes due to the sequence and orientation of amino acids; and the emotional content of music encoded in temporal series of notes.) However, the reductive breakdown of ecosystem services provided by different atmospheric layers, as shown in Table 3.1, highlights a number of important lessons about atmospheric services.

Firstly, these services are diverse. They are also vitally important for the processes of life, just as the processes of life are vital for the structures and functions that produce atmospheric services. They are also highly interdependent: the regulatory shielding service provided by the magnetosphere shields all the inner layers of the atmosphere; the ozone layer in the tropopause screens out biologically harmful radiation; while the tropospheric circulation of pollen and other biological

TABLE 3.1 The services provided by air and the atmosphere

Atmospheric layer	Provisioning services	Regulatory services	Cultural services	Supporting services
Magnetosphere	None identified	Shields cosmic and solar rays Protects the physical structure of the lower atmosphere, including all of its beneficial services	Particles in the solar wind causing aurorae have tourism benefits	None identified
Thermosphere	Outer thermosphere (exosphere): Supports satellites (mainly through centrifugal forces) yet low density does not inhibit momentum Inner thermosphere (ionosphere): Supports communication as radio waves are bounced off the ionic layer	None identified	None identified	None identified
Mesosphere	None identified	Incinerates meteoroids falling to Earth	None identified	None identified
Stratosphere	None identified	Protects from radiation, plasma and meteors[1] Natural global warming of 33°c[1] Cleansing capacity[1] Disperses air pollutants[1]	Aesthetic, spiritual and sensual properties[1]	Redistributes water[1]
Tropopause	Fast transport of water, gases and suspended particles in the jet stream The jet stream has been recognised, though not yet exploited, as a potential source of energy for harvesting	The ozone layer shields the troposphere from damaging ultraviolet and other biologically active radiation	Acceleration and energy conservation of aircraft in the jet stream	None identified

Troposphere	Upper and lower troposphere: Supports ecosystems and agriculture[1] Sound, communications and transport[1] Lower troposphere only: The air that we breathe[1] Combustion of fuel,[1] including gases consumed for this purpose[2] Power generation (wind and wave)[1],[2] Extraction of atmospheric gases[1],[2]	Reflects solar radiation (albedo)[3] Buffers vibrations due to the viscosity of air Both transmits and attenuates sound Insulation (thermal, electrical, etc.) including thermal buffering Connects habitats, including important life stages (pollen, propagules, seeds, etc.)[2] Drying (paint, laundry) Dispersal Weathering (services and disservices) Atmospheric processes that contribute to regulating air quality[2] Climate maintenance (including on global and microclimate scales)[2] Dissipates waste gases, including those produced by combustion processes[2] Conveys and dissipates storm energy[2] Redistributes water, including purification through evaporation and condensation[2] Aeolian erosion[2] Deposits airborne particles (physical and biological)[2] Transports pollen and other propagules[2] Transmits and/or breaks down disease organisms[2] Mist suspended in the troposphere is used in some three-dimensional projection technologies	Contributes 'fresh air' and a sense of open space to the enjoyment of places (recreational, aesthetic, contribution to local character, etc.)[2] Enables sports (balls, shuttlecocks, hang-gliding, etc.)[2] Transport, including both suspension and propulsion (aeroplanes, sailing boats, etc.)[2] Wind/air-related deities, traditions and other folklore[2] Air burials, including *kriya karam* and other funeral ceremonies Artistic quality of light Conveys music, speech and other forms of communication	Heat and water fluxes[3] Flow of water through plants to the atmosphere (evapotranspiration)[3] Recycles gases[2] Recycles nutrients (gaseous and particulate)[2] Recycles water[2] Habitat for wildlife (including for flight and migration)[2] Natural regime of light and darkness (periodicity, spectrum and polarisation) to which species and ecosystems are adapted Natural background of sound, against which species navigate, communicate, hunt or evade, and otherwise adapt behaviours
Whole atmosphere	Broken down by layer above	Broken down by layer above	Atmospheric recreation/tourism,[1] but also including the aurorae identified above as well as astro-tourism, which depends on the clarity and other properties of the whole atmosphere	Note: The overall structure and functioning of the whole atmospheric system provides resilience and supports the production of all other services, analogous to the 90% of an iceberg invisible below the waterline, which, though of substantial value, is also substantially overlooked

Sources: 1. Thornes et al. (2010); 2. Everard et al. (2013); 3. House et al. (2005)

matter, nutrient substances and gases maintains the ecological cycles that regenerate the upper atmosphere's structure and functioning. The whole atmosphere contributes to the cycling of water, heat and dissolved gases in the oceans and in the wider biosphere as well as to earth-forming processes, while the geodynamo of the Earth's outer core generates the protective fields of the magnetosphere.

It is important to recall that all ecosystem services reflect not only the things that ecosystems provide to humanity, but also the interests of diverse sectors of society with their correspondingly diverse value systems. All values, whether currently marketed or not, are valid and important. A non-systemic perspective, often shaped by a narrow financial or other utilitarian focus, can result in services being overlooked, and that can cause problems. This can include the externalisation from decision-making of important processes, including supporting ecosystem services and many of the regulatory services that are vital for maintaining the integrity, resilience and functioning of the airspace and other environmental media. Non-systemic omissions may also result in unsound policies, such as the one that led to public uprisings forcing a UK government U-turn over plans in 2011 to sell off publicly owned forests. In this case, the government had overlooked the array of meanings that these places have for people over and above the narrow financial calculations reflected in their plans.

It is notable in Table 3.1 that the troposphere, the part of the atmosphere in which humans and most of life lives and that is the principal vector of weather and other systems, has the most services assigned to it. By a large margin, most of the recognised services are regulatory in character, reflecting the role of this inner atmospheric layer in connecting, separating and buffering. However, this does not imply that the other layers of the atmosphere are less important, as all provide critical services to maintain the integrity and functioning of the whole, nor does it mean that their full range of services is known or appreciated. All the lists of services within the different cells of the table should be regarded as merely a draft, based as they are on our far from perfect current knowledge – this is particularly true of the blank cells, which therefore warrant further investigation.

The economic value of air

On the basis of the wide range of observed services, the atmosphere is clearly one of the most valuable resources on the planet. Because they are generally invisible, for all their vastness and importance, the air and the atmosphere still tend to be a neglected resource that is largely taken for granted yet is increasingly exploited and commodified.[8] Consequently, human activities have reduced the atmosphere's capacity to supply the services upon which ecosystems and humans depend, degrading essential ecosystem services and overall resilience.

Ecosystem services constitute a systemic framework with which to assess the plurality of societal values flowing from natural assets to humanity, and to begin to assign relative importance or value to each. Assigning economic value to air and the atmosphere is problematic as their services represent irreplaceable life support systems without which there would be no life, let alone people or economic activities. Furthermore, many supporting and cultural services are simply not traded, and so defy traditional monetisation. However, if values are not recognised in some form, they remain external to the economy and, as such, will almost certainly be assigned zero worth and will remain excluded from corporate, governance and other forms of decision-making, with potentially detrimental outcomes. Valuation in some form or another is therefore essential, although economics and accountancy are different things: the former relates to an assessment of the relative value of something to society, whereas the latter takes a more simplistic approach to monetising and aggregating tradable benefits and costs.

On the basis of the subset of services that they identified, John Thornes and his colleagues estimated a total economic value of the atmosphere of between 100 and 1,000 times the gross world product, which stood at approximately £43 trillion in 2008. Expanding upon these figures in a conference presentation in September 2011, Thornes proposed that the 'value of air' is approximately 1p (UK pence) per cubic metre, using the then current price for carbon dioxide within the EU emissions trading system as a basis.[9] Based on this calculation, each person in the UK 'costs' £65 per year for their share of the global atmosphere for breathing alone, or around one-seventh of what the average person spends annually on gas and electricity.

As other services are harder to value in robust terms, and quantification of damage to them from a range of negative human impacts is elusive, the 'real' value of a stable and functional air and atmospheric system is vastly more than this. A further problem also arises from the sheer scale of the atmosphere. On the one hand, this may mean that some impacts, such as the extraction of gases for industrial, medical, agricultural and other uses (provisioning services), may be minimal or negligible, though unquantifiable. On the other hand, chronic negative effects such as ozone depletion or climate change can manifest themselves only slowly, and thereby escape detection, valuation or concern until their effects become significant. Impacts may also be remote in time and space from the pressures that erode the service, sometimes with poor or no understanding of 'cause and effect' pathways.

Notwithstanding the difficulties with quantification and valuation methods, all atmospheric services are essential for continuing human wellbeing. There is therefore a pressing challenge to recognise and value these important services in decision-making, backing them up with robust valuation, legislation and common law precedents, if we are to direct society onto a more sustainable course.

Nevertheless, air and its services are still largely assumed as being 'free', with virtually all of the less tangible services omitted from general planning considerations or markets. Consequently, they remain subject to continued degradation through incautious overexploitation, blind or ambivalent to the consequences for the beneficiaries of other ecosystem services. This is a classic repetition of Garrett Hardin's 'Tragedy of the commons'.[10]

All negative effects, however slow to manifest, may be individually and/or cumulatively significant, suggesting that a far more considered and precautionary approach is required to reflect on the wider ramifications, and the associated benefits and costs, of all our uses and abuses of the airspace.

4 | Abuses of the air

The last three chapters have celebrated respectively our deep biophysical interdependence with the breathing space we inhabit and how it has shaped society, and the multiple if often formerly overlooked ecosystem services that it provides. This chapter turns to what happens in the reverse direction. Beyond the consequences of our basic biological interdependence, what are the other, less favourable legacies of societal activities for the air and all of life that share it?

What goes around comes around

Nevil Shute's chilling 1957 novel *On the Beach* tells a post-apocalyptic tale of the end of the world, seen through the eyes of a submarine crew as a radioactive cloud spreads relentlessly around the world following nuclear war in the northern hemisphere. Although the fatal fallout derives from the bombs themselves, all the nuclear nations having launched their entire arsenals at each other as a kind of death rattle of mutually assured destruction, it is the winds and weather systems that spread it at a painfully slow pace across the globe, where it is taken up with lethal effect through lungs, water and food. The book's title is taken from a line in the T. S. Eliot poem *The Hollow Men*, and many editions of *On the Beach* quote the poem's concluding lines: 'This is the way the world ends. Not with a bang but a whimper.'

Mercifully fictional, though nonetheless truly disturbing, *On the Beach* exemplifies the finite capacities of the airspace to absorb and assimilate pollution, and its role in distributing all aerial emissions. These emissions include substances that are natural yet problematic in excess, such as oxygen and carbon dioxide exchanged in photosynthetic and respiratory processes, or nutrient substances that, although essential for growth, fundamentally change ecosystem composition and functioning when present beyond limiting concentrations.

The atmosphere receives diverse anthropogenic emissions of destructive substances. These include synthetic chemicals, some of them climate-active or ozone-depleting, as well as bioaerosols, metal-enriched dust and poisons of various other forms. We might have managed much of our post-industrial pollution control regimes implicitly under the maxim that 'the solution to pollution is dilution'. However, this has been revealed as unsafe where pollutants are persistent or emitted in concentrations swamping natural assimilation rates, or where aspects of the atmospheric system itself become damaged by the emissions.

Archaeologists are intensely interested in middens left behind by past civilisations, using artefacts dumped in these waste pits to deduce how ancient people lived. We have treated the atmosphere as just such a waste dump; whether this has been wilful or indirect, we have always incautiously disposed of all sorts of gaseous, particulate, aerosol and other waste substances, so that the air itself carries the signature of modern society. Just as James Lovelock deduced that an unstable gaseous composition in the atmosphere was suggestive of the presence of life, so too the loads of toxic and other damaging substances in the Earth's modern atmosphere are a fingerprint of our cavalier historical pathway of development. This chapter explores the many abuses that humanity has inflicted upon the life-support processes of air, and some of their immediate, medium-term and longer-term consequences.

Spreading a bad odour

Humans, it seems, are programmed to be repelled by, among other things, foul odours. Disgust was recognised as a basic emotion as long ago as 1872 in the book *The Expression of the Emotions in Man and Animals* by Charles Darwin, who recognised the emotion of disgust in human reactions to things that appeared to be revolting. Disgust is also one of the eight emotions within the psycho-evolutionary theory developed by the American psychologist Robert Plutchik (1927–2006), an influential classification of biologically primitive 'basic' emotions that had evolved for their survival value and contribution to reproductive success.[1] Disgust is typically associated with things regarded as unclean, inedible, infectious, gory or otherwise offensive. Some aspects of disgust are learned, with a role in cultural morals, while others appear to be innate. Disgust reactions are particularly strongly associated with the sense of taste, but also with the senses of smell, vision and touch, and it is widely believed that disgust is an instinct that evolved to help us avoid contaminating our bodies. As a primitive emotion, disgust can be a powerful agent of awareness-raising and the mobilisation of people to respond, including to some of the 'stinks' generated by modern society. We are, it seems, culturally as well as biologically sensitive to the contamination of the air.

The 'Great Stink' of 1858 in London, when the River Thames became so polluted that the House of Commons resolved to tackle gross contamination by sewage influx from the burgeoning city, happened at a time when diseases were thought to be distributed by a 'miasma'. This, let us recall, was in the days before the now almost ubiquitously accepted 'germ theory' of disease, which gained traction throughout the nineteenth century. Under the miasma, or miasmatic, theory, it was believed that contagious diseases, including cholera and the Black Death, were spread by a *miasma*: quite literally a form of 'bad air', as the term 'miasma' derives from an ancient Greek word meaning 'pollution'. A miasma was believed to be a poisonous vapour or mist containing particles arising from decomposed matter (miasmata), which were responsible for causing illnesses. It was believed that miasmas were identifiable by their foul smell; marshes and other wetlands were particularly reviled in some cultures due to the sulphurous and other smells emanating from what we now know to be innocuous and beneficial anaerobic processes, but which were then considered likely to waft diseases into neighbouring towns and villages.

There is a direct parallel here with the disease malaria: in mediaeval Italian, the word literally translated as 'bad air'. It was so named before any link between the disease and the insect vectors of the *Plasmodium* parasite was suspected. Modern understandings of the mosquito vectors of malaria broke down the ancient belief that swamp gases contributed to outbreaks of the disease, although a fear of marshes and wetlands remains deeply rooted in different cultures due to their odours of decay.

Cleansing and other hygienic practices thought to work against the effects of the miasma – for example as practised and promoted by Florence Nightingale, who did much to make hospitals sanitary and fresh-smelling – would also have been effective as a means of limiting some of the transmission pathways of microbial pathogens. What the miasma theory and its widespread acceptance did achieve was a recognition that diseases are often the product of environmental factors such as contaminated water and air, exacerbated by poor hygiene. The concerns of the members of the British Houses of Parliament in 1858 may therefore have been more than merely aesthetic; they were also worried about personal and urban health.

Fouling the industrial and urban airspace

As industry boomed and European cities flourished, people were drawn to them both to find employment in the new industrial age as well as to abandon the land as mechanisation displaced farm labour. The urban population of Britain leapt from 2 million in 1800 to 20 million at the turn of the twentieth century; by 1850 in London, the biggest city the world had ever seen, these enormous concentrations of people posed brand new problems of feeding, watering and housing the masses.[2] Factories, transport, domestic fires and all manner of other sources generated increasing loads of anthropogenic pollutants, including gases, aerosols, smoke, ash, soot and dust from combustion, transport, domestic and industrial processes, discharged into urban airspaces that were often confined by ever taller buildings. Consequently, pollutants increased in concentration, with profound effects. These were later to be augmented by mist from aerosol spray cans, cigarette smoke, volatile solvents from paints, and a complex soup of gaseous wastes.

Anthropogenic aerosols have a range of impacts on the climate and on human health. 'Smog' is a familiar term, formed by an amalgam of the words 'smoke' and 'fog', generally attributed to Dr Henry Antoine Des Voeux and his 1905 paper, 'Fog and smoke', that he presented at a meeting of the Public Health Congress. The constitution of smog is as diverse as the sources that create it. In early cities, high densities of coal burning were a feature of smog, which was created from a mixture of smoke, sulphur dioxide and other components, all of which tended to react with each other in a kind of toxic soup. A similar type of smog can be formed by erupting volcanoes, which emit high concentrations of sulphur dioxide and particulate matter; this is generally referred to as 'vog' to distinguish it as a natural and generally more short-lived occurrence. However, the smog in our modern cities, particularly since the 1950s, derives substantially from vehicular emissions and industrial fumes as well as from coal fires, which combine and react with sunlight to form a range of secondary pollutants. This is known as 'photochemical smog', which today affects nearly every large city on Earth.

During the Victorian era, London was notorious for its thick smogs, known as 'pea-soupers'. In fact, concerns about London's air date from long before that. Indeed, in 1306, fears about air pollution were sufficient for King Edward I to institute a ban on the burning of sea coal. Smog episodes persisted throughout the nineteenth and twentieth centuries, particularly during the winter when more house fires were burning coal. Things came to a head in 1952 with the 'Great Smog', which blanketed London's streets for four days, killing an estimated 4,000 people. It is estimated that a further 8,000 died from its effects in the following weeks and months,[3] although this was initially dismissed as an influenza epidemic. This in turn led the government to institute the Clean Air Act 1956, which introduced smokeless zones into the capital. Resulting declines in concentrations of sulphur dioxide consigned to history London's intense and persistent coal-derived smog episodes. However, smog and other pollution problems have changed in nature with technological progress, giving rise to contemporary issues associated with petrochemical smog.

Petrochemical smog comprises a complex mixture of nitrogen oxides and a wide range of volatile organic compounds, both types emitted largely as a result of the burning of fossil fuels. Once released into the atmosphere, these substances react with sunlight to form many noxious compounds, including carbon monoxide, particulate matter, ground-level (or tropospheric) ozone, sulphur dioxide and nitrogen dioxide. These are all implicated in respiratory disorders and premature deaths, and some also with stimulating cancers. Senior citizens, children and people with respiratory and heart conditions are particularly vulnerable, with hospital admissions and deaths often rising during periods of hot and stable sunny weather when tropospheric ozone levels are high and layers of air at higher elevations are warm enough to inhibit vertical circulation.

Fine particulates, particularly $PM_{2.5}$ and PM_{10} (particles smaller than 2.5 micrometres and 10 micrometres respectively), have emerged as a serious threat to human health. $PM_{2.5}$ is of particular concern as this represents the respirable fraction, reaching deep into the alveoli of the lungs. This fine matter can comprise or adsorb problematic substances that can rapidly be conveyed into the human body. *Baseline Scenarios for the Clean Air for Europe (CAFE) Programme* notes that:

> New studies show that exposure to small particles (below a diameter of 2.5 μm [micrometres], $PM_{2.5}$) is associated with substantially increased mortality, especially from cardio-vascular and cardio-pulmonary diseases. Present levels of $PM_{2.5}$ in Europe are now estimated to reduce the statistical life expectancy in European population by approximately nine months, comparable to the impacts of traffic accidents. Thus, these newly identified impacts of fine particles by far exceed those identified earlier for ozone.[4]

As yet, no concerted response to this threat has been formulated.

A wide range of modern societal activities also release or generate a varied range of substances, some synthetic and others naturally occurring but now in elevated concentrations. These substances have a propensity to accumulate in the environment and may reach unsuspected trigger concentrations, resulting in environmental and human health impacts. One among many such substances of concern is a group of chemically related persistent compounds known as dioxins. Although dioxins can and do result from natural processes such as forest fires and volcanic eruptions, which fuse chlorine atoms with organic matter, the bulk of the current environmental burden is produced as a by-product of industrial processes including smelting, chlorine bleaching, herbicide and pesticide manufacture, as well as uncontrolled waste incinerators and emissions from internal combustion engines. Aerial dispersion is a key environmental pathway, although only very low levels of dioxins are found in the air itself. However, these substances are present in environments across the world, at their highest levels in some soils, sediments and food, especially dairy products, meat, fish and shellfish.[5] More than 90% of human exposure to dioxins is through food, and these dioxins contribute to reproductive and developmental problems, damage to the immune system, interference with hormones and also the stimulation of cancer. Dioxins can be destroyed, and creation of dioxins in waste gases prevented, by combustion at temperatures in excess of 850°C; clearly, incautious emissions into the air have to be controlled carefully if serious unintended problems are to be averted.

Pollution of urban spaces with health-related gases and smog can be exacerbated by topography and dense populations. Cities such as Mexico City, Santiago, Tehran, Los Angeles, Singapore and San Francisco, and rapidly expanding Chinese cities such as Beijing and Shanghai, are particularly prone to smog as their dense buildings trap air, causing smog levels to rise along with their associated health and wider economic issues. The human health implications are serious: for example, China's former health minister Chen Zhu claimed that China's 'airpocalypse' accounts for between 350,000 and 500,000 premature human deaths each year (equivalent to the population of the British city of Bristol), with air pollution representing 'the fourth biggest threat to the health of Chinese people' (after heart disease, dietary risk and smoking).[6] The European 'Year of Air' was 2013, at the end of which, in December, the European Commission announced the Clean Air Policy Package[7] to clean up the continent's air by 2030. Proposals include national emission ceilings for a range of health-relevant air pollutants, which will contribute to progressive declines in pollutant concentrations. Welcoming the initiative in principle, an editorial in the authoritative medical journal *The Lancet*[8] called for more urgent action and stringent standards; standards for $PM_{2.5}$ in particular are considered especially lax given emerging knowledge about their link with heart attack risk.[9]

Seeing the wood for the trees

As we observed in Chapter 1: *Air and the making of the atmosphere*, trees, and particularly large forests, have a major role to play in connecting earth, water and air, including weather and climate systems and cycles of water and carbon. Research on the carbon pools and flux associated with global forest systems found that, globally, forest vegetation and soils contain about 1,146 petagrams (10^{15} grams) of carbon, over two-thirds of which relates to soils and associated peat deposits.[10]

It should come as no surprise, then, that human activities impinging on trees have a massive impact on the cycles, stability and services of the atmosphere. The impacts of human deforestation, reforestation and afforestation are profound. Deforestation worldwide is responsible for up to 20% of global greenhouse gas emissions.[11] The clearance of forests, particularly well documented in the media in the Amazon and in Indonesia, raises major concerns for the stability of climatic systems on a global scale. Worryingly, there is more than a suspicion of positive feedback: the loss of tree cover is directly driving climate change, but the changing climate itself is now affecting the Amazon. There are concerns that more than half of the Amazon rainforest could be lost or severely damaged as early as 2030 if current trends in deforestation, droughts, forest fires and global greenhouse gas emissions continue. These concerns are, however, far from localised. The continuing loss of the Amazon's rainforests can lead to droughts and crop failures not merely in the Amazonian region but, through the forests' profound influence on climatic and airborne water flows, it may contribute to similar problems as far afield as the grain belts of South America and North America and possibly in other agricultural regions as distant, yet still tropospherically interconnected, as Europe.

Forest cover in mountain regions can be especially significant in trapping moisture that may then be recirculated later to run downwards in buffered flows to water entire continents. One such forest-covered ridge is the Western Ghats, running down the western seaboard of India's Deccan peninsula, which intercepts moisture from warm winds blowing in from the Arabian Sea. The Western Ghats not only intercept this water but its ecosystems, particularly its matrix of wet and dry forests, store and recycle it, releasing it slowly into the three great rivers that flow eastwards across the entirety of southern India to irrigate land that would otherwise be arid. Forest clearance on the Western Ghats is giving rise to multiple concerns, including the loss of both endemic biodiversity and the close connection between moist air streams and the water, land and livelihoods of millions of people.[12] If forest loss continues, there is a risk not merely that this connection will be degraded, but that hot air from bare mountains will deflect the warm winds higher, robbing peninsular India of its life-giving sustenance of wind-borne water, with devastating and possibly permanent consequences.

The loss of trees and other forms of green space in urban settings contributes to such problems as the development of 'heat islands' and the perpetuation of air pollution. Conversely, as we will see in Chapter 6: *Thinking in a connected way*, the reinstatement of urban trees can play a major and cost-effective role in restoring these atmospheric services.

Downsides of land use

It is not merely direct emissions into the airspace that generate the kind of atmospheric pollution problems that we have touched upon in this chapter. So too do the ways in which we manage or abuse other environmental media. For example, during the 1960s, fisheries on the Aral Sea, then the world's fourth largest inland sea, supported a huge Soviet cannery industry on the shoreline. At the same time, Soviet engineers were tapping ever more substantially into the Syr Darya and Amu Darya rivers that fed the Aral Sea in order to increase the productivity of a cotton industry that had been present for decades. The two initiatives were not mutually compatible; the last trawler was abandoned in 1984, along with the canneries, as the Aral Sea receded, starved of fresh water and dried up by the tropical sun. By 2003, continuing declines in the water level saw the Aral Sea not only shrink to a quarter of its former area, but also separate into two small hyper-saline seas. Today, the city of Aralsk stands not as it once did on the seashore, but instead 40 miles (64 kilometres) north of it, separated by a huge and unproductive salt pan. However, this is far from being the full extent of the environmental, social and economic damage. Now, winds whip up a dangerous cargo of salt and pesticide residues from arid former lake bed surfaces, destroying the ecology, economy and health of central Asia.

Contamination of the air with dust is a far more widespread problem. The simple word 'dust' hides a huge diversity of particle types that may enter the atmosphere. Some of these dust particles are formed from natural matter born in the Big Bang, others are of biological origin, and a wide range has a human origin. Dust particles from natural sources enter the atmosphere through volcanic eruptions and aeolian processes, as well as in the form of pollen and other biological substances. Natural atmospheric dust transport is a significant contributor to earth-forming processes and, to some extent, to nutrient flows – for example, Saharan dust is transported and deposited as far away as the Caribbean and Amazonia – as well as influencing air temperatures and rainfall.[13] Dust deriving from human sources can include incompletely burnt carbon particles, fibres from sources as diverse as textiles and paper, skin cells and the products of quarrying. Indoor dust can be particularly dense, and includes house mites and their faeces. All of these substances can be allergenic, contributing to problems such as hay fever.

The loss of topsoil due to wind and water erosion now exceeds the natural formation of new soil over large areas of the world. This progressively robs the land of its topsoil and fecundity, raising concerns about food supplies, which must necessarily increase to feed the planet's burgeoning human population and to meet the demands of its rising middle class. Sediment whipped up by the wind is conveyed on air currents, to be redistributed elsewhere; not only does this contribute to geomorphological processes but also, when it lands in populated areas and on built infrastructure, it is both unwelcome and potentially harmful and expensive to manage. We are, in fact, witnessing increasingly large dust storms around the world: for example, huge dust plumes now commonly shade out the sun above the cities of north-east China. In early 2014, storm-driven Saharan dust spreading across southern Britain triggered a serious health warning for asthmatics and those with heart and chest complaints; the dust had the potential to stimulate asthma in undiagnosed sufferers, with symptoms persisting perhaps for several weeks after the dust problem had abated.[14] Eroded matter from the land also contributes to higher-altitude aerosols, which enter and disperse into the atmosphere and have potential implications for energy exchange.

Agriculture is also a significant source of ammonia emissions into the air. Estimates for total UK ammonia emissions in 2006 stood at 315×10^3 tonnes per year, of which 288×10^3 tonnes per year, 90% of the total, originated from agriculture. Most of this occurs after the spreading of livestock waste, with livestock housing and grazing accounting for much of the rest and a smaller contribution from nitrogen fertilisers and crop production. Ammonia is problematic due to its impacts on human health, but also for a wide range of additional reasons including the eutrophication and acidification of non-agricultural soils and water, as well as representing a loss of valuable farm nutrient resources.

We will explore issues of climate change shortly, but highlight here that agricultural activities are also responsible for generating a range of greenhouse gases, including carbon dioxide but also methane and nitrous oxide, which are respectively 21 and 310 times more potent than carbon dioxide in terms of radiative forcing (the difference between energy absorbed by the Earth from sunlight and energy radiated back into space) over a 100-year period. The food production cycle also uses and gives rise to significant consequent but unintended releases of additional exotic substances, including refrigerants such as chlorofluorocarbons, which are used primarily in food production processes. In addition, farms are the cause of a wide range of odour complaints: statistics compiled and summarised by UK environmental health departments between 1987 and 1990 record that 44% of complaints related to manure and slurry spreading, 25% to livestock buildings, and 21% to manure and slurry stores. Pig farms elicited the highest number of complaints. Given that 70% of the land area of the UK is farmed, a proportion that is representative of the extent of agricultural land use worldwide, the contribution of farming to a wide range of air-related pollution issues must not be overlooked. Land use can also significantly change the physical and biological properties of the land surface, with implications for the climate system, dust and air pollutant generation, and other regulatory attributes of the atmosphere. While substantial efforts are being made to address some of these impacts, the goal of controlling atmospheric contamination and farming sustainably remains challenging.

The role of wetlands

Wetland systems also play a major role in connecting earth, water and air, including in carbon budgets, nutrient cycling, water exchange and impacts on the climate on both localised and wider scales. One appraisal of global wetland areas and their organic carbon stock found that, through their wealth of stored carbon, wetlands provide a major potential sink for atmospheric carbon; on the other hand, if they are not managed properly, they could become substantial sources of greenhouse gases such as carbon dioxide and methane.[15] So the massive scale of global wetland loss, largely at the hands of agriculture, has major implications for the supply of a wide range of beneficial ecosystem services.[16]

The importance of many of these ecosystems services has been overlooked in the past; localised benefits are still promoting their overuse and degradation, despite the net cost to other service beneficiaries, from catchment to global scales. So wetland conservation, and the valuation of wetland ecosystem services, is strategically important for the integrated protection of the resilience and services of air and the atmosphere.

However, not all natural water–air interactions are benign for wildlife and people. For example, the mass mortality of local people, stock and wildlife has been associated with various tropical lakes as a result of 'limnic eruption'. This phenomenon, also known as 'lake overturn', is a rare form of natural disaster in which a very high concentration of dissolved carbon dioxide that has accumulated in a highly stable deep lake suddenly erupts, much like a fizzy drinks bottle effervescing when pressure is released. Examples of this rare occurrence include Lake Monoun in 1984 and Lake Nyos in 1986, both in Cameroon, where supersaturated carbon dioxide originating either from volcanoes or from the decomposition of organic matter was released. The trigger setting off the limnic eruption in Lake Nyos was thought to be a landslide, but other triggers could include volcanic eruptions, earthquakes, or extreme wind or rain storms. A substantial cloud of carbon dioxide forms above the lake following the eruption then expands across the neighbouring region, and, due to its higher density, it displaces the breathable air.

Limnic eruption is, however, a very rare phenomenon, and, notwithstanding its devastating and lethal local effects, it has little impact overall on the global atmosphere. But it does emphasise that there are many, often significant and overlooked links between wetlands, water systems and the atmosphere, which are therefore intimately and constantly interacting.

Other indirect contributions

Many other facets of the ways in which we use resources and technology in modern life pose indirect threats to the air system. The many waste gases that can be produced by landfill sites are just one example. Ostensibly a means to dispose of solid waste 'out of sight and out of mind', the burial of waste in various tips or dumping grounds leads to a host of problems that have become increasingly evident due to the increasing volume and complexity of waste materials.

One of the major concerns is pollution, including the seepage of a complex cocktail of liquids that contaminate aquifers and soil. However, the off-gassing of methane generated by decaying organic matter is one of the more harmful emissions to the atmosphere, where, as we will see, it makes a significant contribution to climate change. Additional wider ramifications of landfill sites include not only the noise generated by landfilling operations but also the fact that they attract and harbour potential pests and disease vectors such as rats, flies and gulls, as well as generating aerial nuisances such as dust and odour.

Reversing evolution

Chapter 1: *Air and the making of the atmosphere* highlighted the progressive purification of the atmosphere from the homogenous cloud of space dust through to the onset of the industrial age. Since that time, society's heavy dependence on mined substances and their cavalier disposal have resulted in the systematic accumulation of formerly sequestered substances in the biosphere, including the venting of waste gases into the air.

Deposits of fossil carbon, metals, phosphorus and other substances now mined to support modern lifestyles are a product of progressive sequestration from the atmosphere and the wider biosphere over geological timescales. To release these mined substances back into the biosphere is therefore inherently dangerous, as accumulating concentrations in the air reflect earlier, more contaminated phases of biospheric history. This effectively returns the environment to a prior condition that supported simpler forms of life but was less suited to the more complex life forms that co-evolved with decreasing pollutant levels. Once emitted, these persistent substances will not simply 'disappear' but will tend to disperse into natural cycles. As we know from surveys of river invertebrates, pollution tends to result in the loss of more complex creatures but allows simpler organisms to thrive, and the same principle applies more generally in the biosphere. For some commentators, the progressive return of historically sequestered pollutants into the biosphere is analogous with 'reversing evolution'.

This principle also applies to incautious releases of synthetic substances that cellular enzyme systems and other natural systems have not evolved to break down. Examples include a variety of man-made chemicals, as well as often uncharacterised substances resulting from combustion, such as gaseous and PM_{10}, $PM_{2.5}$ and other particulate matter and their adsorbed pollutants. These may be taken up by cells as organic matter, but, as the cells lack enzyme pathways to break them down, they tend to accumulate there with unknown effects. In the atmosphere, they can both persist and carry breakdown products into different compartments of the wider atmosphere; as we will see later in this chapter, these products can include long-lived climate-active gases and halogen free radicals that are implicated in the breakdown of stratospheric ozone.

Complex substances, complex interactions

It is clear from the progression of different types of smog, the contribution of man-made substances to emerging problems, and the diversity of issues derived from agriculture that we are dealing not only with complex problems but also with complex environmental interactions. If we assume that there are only simple cause-and-effect pathways and equally simple solutions, we underestimate the complexity of air pollution. Take, for example, elevated ground-level ozone, implicated in significant health concerns and in environmental consequences such as the suppression of crop productivity.

A range of human health impacts are associated with exposure to elevated levels of ground-level ozone. These include impacts on lung function and inflammation of airways as well as respiratory symptoms including coughing and irritation of the throat, and tightness, pain and discomfort in the chest. More seriously, there are links between higher daily exposure to ozone and increased asthma attacks, reduced immune response at the lung surface, stimulation of cancers, and more frequent visits to doctors and hospital admissions. Ozone exposure is associated with increased mortality – considerably so for older adults and during warm periods of weather. In common with many urban air pollutants, elevated levels within microenvironments give particular cause for concern, for example adjacent to busy roads, in areas enclosed by buildings and inside vehicles and homes. Perversely, the additional volume of air and its penetration into deeper areas of the respiratory tract may exacerbate health risks for people exercising heavily in areas with higher ozone levels.

Elevated tropospheric ozone concentrations also have significant effects on vegetation. Impacts on natural vegetation and ecosystem functioning are less well studied, perhaps indicating a major knowledge gap. However, the magnitude of the consequences for crops is giving rise to food security concerns.[17] While plants are able to detoxify low concentrations of ozone, beyond this level ozone absorbed through the leaves is implicated in a range of impacts, including yellowing of the leaves, premature leaf loss, decreased seed production and reduced root growth, cumulatively reducing yield quantity and/or quality as well as lowering resilience to other stresses such as drought and disease. Through these mechanisms, ozone pollution causes millions of tonnes of crop losses each year, and not just in the regions where the air pollutants are emitted but across continents. Two of the world's most important staple food crops, wheat and soya bean, are particularly sensitive to ozone pollution, with an 18% reduction in yield shown in laboratory experiments, and there is an observed 10% reduction in rice, maize and potatoes.[18] This has major implications for global crop production. Transboundary pollution is of particular concern, with ozone emissions from North America reducing wheat yields in Europe by 1.2 million tonnes each year and, on a global scale, pollution from South East Asia causing the loss of 6.7 million tonnes of wheat and about 11.6 million tonnes of rice each year.[19]

Yet, at higher atmospheric altitudes, it is precisely the loss of ozone that is at the root of serious environmental and health threats. As outlined in Chapter 1: *Air and the making of the atmosphere*, the ozone layer in the lower stratosphere serves as a shield that absorbs and deflects from the Earth's surface ultraviolet radiation (particularly energetic UVB radiation with a wavelength of 280–315 nanometres) and other ionising radiation emanating largely from the sun. UVB in particular is linked with biological consequences including skin cancers (particularly basal and squamous cell carcinomas and malignant melanoma) and cataracts in humans and other animals, damage to both natural and crop plants, and reduction of oceanic plankton populations. Statistics demonstrating a causal link between ozone depletion and human skin cancer and eye damage are elusive, in part because UVA (radiation with a wavelength of 400–315 nanometres), which is implicated in some forms of skin cancer, is not absorbed by ozone, but also because it is nearly impossible to control for lifestyle changes and due to a lack of reliable historical (pre-ozone hole) surface ultraviolet data.

The thinning of the ozone layer due to the catalytic activity of certain categories of synthetic, persistent chemical compounds has been the source of substantial concern as it is resulting in increasing amounts of ionising radiation reaching the Earth's surface. Monitoring of this layer throughout the latter half of the twentieth century has revealed an alarming reduction, particularly around the South Pole, where in some parts the thinning is extreme enough to be defined as 'holes'. In fact, the term 'ozone depletion' describes two related though distinct phenomena that have become increasingly evident and well understood since the late 1970s. The first of these is a net decrease of around 4% per decade in the total volume of stratospheric ozone. Declines vary with latitude, from about 3% below pre-1980 values between 35 and 60 degrees north to about 6% for 35 to 60 degrees south, while no significant trends have been observed in the tropics.[20] The second phenomenon, referred to as the 'ozone hole', relates to the more substantial springtime decrease in stratospheric ozone over the polar regions. Reductions of up to 70% in Antarctica's ozone column observed in springtime were first reported in 1985.[21] In the Arctic, ozone losses are more variable year to year, rising to 30% in the winter and spring when stratospheric temperatures are colder.

For the two linked phenomena, the causative agent is the same: catalytic dissociation of ozone by atomic halogen elements, principally chlorine and fluorine. Naturally, only trace concentration of halogens would be present as high up as the stratosphere. However, durable man-made halocarbons, and particularly refrigerants such as chlorofluorocarbons (CFCs), other freons and halons (alkanes with linked halogen elements) can persist in the atmosphere, carrying with them their halogen constituents. While CFCs admirably served as air conditioning and refrigeration chemicals, aerosol propellants and cleaning solvents for delicate electronic equipment due to their stability and lack of electrical conductivity, the fact that they do not occur naturally means that natural systems, such as cellular enzymes, have not evolved to break them down, and this durability means that they persist in the atmosphere. After being emitted at the surface, they enter the atmosphere as volatile substances and they may then circulate freely, some eventually reaching the stratosphere. Once there, their chemical constituents become dissociated by intense incoming radiation, releasing free halogens that proceed to catalyse ozone degradation. Some other free radicals, including hydroxyl radical ($OH^{\bullet}$) and nitric oxide radical ($NO^{\bullet}$), can also be released by man-made chemicals with ozone-depleting effects. CFCs and other substances catalysing the breakdown of stratospheric ozone are generally referred to as ozone-depleting substances (ODSs). Some ODSs, such as chlorine free radicals, can remain active in the stratosphere for a number of years before chemical processes finally remove them into lower atmospheric layers. These ozone-depleting impacts are of great concern due to the vital role of the ozone layer in shielding the Earth's surface from biologically damaging radiation.

Nutrient enrichment vectored through the atmosphere is also a pressing problem. The massive scale of mining of nitrogenous fertilisers from the abundant store of largely biologically unavailable nitrogen gas in the air was outlined in Chapter 2: *Living in a bubble*; about 30% of the total fixed nitrogen on Earth is now manufactured in ammonia chemical plants.[22] There is no evidence that extraction leads to depletion of the nitrogen stock of the air, largely because of the sheer volume and free mixing of the atmosphere. However, a range of problems result from the 'downstream' uses of nitrogen in fixed and biologically available forms. Substantial concerns arising from its widespread application in agriculture relate, for example, to the eutrophication of soils and water and the generation of nitrous oxide and ammonia emissions to the air. Agriculture is estimated to be responsible for over 60% of total emissions of reactive nitrogen in the UK, and the figure is similar in other northern European countries.[23] Human activities such as fossil fuel combustion, the use of artificial nitrogen fertilisers and the release of nitrogen in wastewater have dramatically altered the global nitrogen cycle, affecting ecosystem structure and function as well as the many services that people can expect from closely connected atmospheric, land and water systems. The release of biologically reactive nitrogen into the environment and the perturbation of the planet's nitrogen cycle represent a global-scale experiment that is second only in perceived risk to that associated with emissions of carbon dioxide.[24]

Winds of change

Carbon dioxide, water vapour and other atmospheric gases play an important natural role in making this planet habitable through the 'greenhouse effect'. However, while natural cycles maintain the concentration of greenhouse gases, stabilising the atmosphere and ambient temperatures, atmospheric concentrations of a range of greenhouse gases today are rising sharply.

All organisms influence their environment, and in this regard humans are no different. However, where humanity departs is in the scale of our activities. Particularly since the European Industrial Revolution, commencing in the mid-eighteenth century, the scale of human impact has increased from the local to the continental and the global. Industrial and domestic activities generating greenhouse gases, particularly the combustion of fossil reserves that have sequestered carbon from the atmosphere over geological timescales, have had a profound effect on the concentrations of gases and aerosols within the airspace. In addition, releases of carbon formerly stored in soils, forests and wetlands that have been converted for agriculture and other activities have increased the atmospheric load of climate-active gases.

Prior to the start of the Industrial Revolution in the latter half of the eighteenth century, carbon dioxide emissions from the burning of fossil fuels were negligible, and the atmospheric concentration was estimated at 280 parts per million (ppm) – a figure that had remained relatively constant for about a thousand years.[25] However, the concentration of various greenhouse gases has increased significantly to more than 30% above pre-industrial levels, and is increasing on average 0.4% per year due mainly to fossil fuel combustion and deforestation. By 1950, global emissions of carbon dioxide from fossil fuel burning had reached 1.6 billion tonnes per year, a quantity that was already boosting the atmospheric carbon dioxide level. In 2000, emissions totalled 6.3 billion tonnes, increasing mean atmospheric concentrations to 370 ppm; this represented a rise of 32% from pre-industrial levels, and the rate of increase was estimated at 0.5% per year and rising. The build-up of atmospheric carbon dioxide from 1960 to 2000 of 54 ppm far exceeded the 36 ppm rise from 1760 to 1960. According to the annual greenhouse gas bulletin published by the World Meteorological Organization (WMO) in November 2013,[26] there were 393.1 ppm of carbon dioxide in the atmosphere in 2012, an increase of 2.2 ppm from 2011. Atmospheric levels have risen each year since annual measurements began in 1959, making this one of the most predictable of all environmental trends. The concentrations of other greenhouse gases also give cause for concern. CFC levels are rising by 5% per year, and nitrous oxide levels by 0.4% per year. The ongoing situation is complex; it is estimated that the global rate of increase fell below this growth rate in 2008 due largely to high oil prices in the first half of the year and the economic slowdown in the second half, although increasing biofuel production also helped displace a substantial volume of fossil fuel petrol and diesel. Nevertheless, the developing world, which is seeing the world's major growth of greenhouse gas emissions due to burgeoning populations, increasing per capita material demands and implementation of dirtier, often outmoded technologies, accounted for 50.3% of global emissions in 2008, exceeding the combined emissions of the developed world and international travel for the first time. In this complex and uneven situation, with demand for fossil fuels rising once again as the global recession eases, strong and concerted international leadership and action are required if we are to avert serious consequences from unconstrained economic activities.

The WMO was established in 1950 as an intergovernmental organisation, although it originated in the International Meteorological Organization, which was founded in 1873. It is the specialised agency of the United Nations (UN) for a range of meteorological issues. The WMO annual greenhouse gas bulletin found that levels of gases in the atmosphere that drive global warming had increased to a record high in 2012.[27] The concentration of carbon dioxide grew more rapidly than its average rise over the past decade, while those of methane and nitrous oxide also broke previous records. The WMO concluded that the climatic warming effect had increased by almost a third since 1990, building upon the 141% increase in global average carbon dioxide concentrations in the atmosphere since the start of the industrial era around 1750.

Given the close interdependence of all life, including humanity, with the climate system, human-induced climate change is a matter of grave and rising concern. Carbon dioxide emitted by human activities remains in the atmosphere, beyond the fraction absorbed by plants, trees, the land and the oceans. The 1995 Intergovernmental Panel on Climate Change (IPCC)[28] Second Assessment Report[29] presented a compelling body of scientific evidence supporting the postulate that human activities are already influencing the climate by changing the composition of the atmosphere. This conclusion was reinforced by the 2007 IPCC Fourth Assessment Report,[30] which found with 90% confidence that human activities were the principal cause of 'Most of the observed increase in global average temperatures since the mid-20th century'. The 2013 IPCC Fifth Assessment Report[31] not only increased the confidence of this conclusion to 95%, but also found that human activities, not natural variations, were the cause of all of the global warming observed from multiple and consistent records since 1951. Indeed, it is possible that anthropogenic emissions would drive far greater global warming were their impacts not moderated by natural regulatory processes. The 2013 IPCC statement addressed all human influences on the climate, including not only greenhouse gas emissions but also human aerosol emissions that scatter sunlight and may therefore have a cooling effect which to some degree offsets rising temperatures due to increased climate-active gas concentrations.

A claim of 95% certainty, as a consensus of the many scientists globally studying this phenomenon, is about as strong a statement of certainty as the scientific community makes. The IPCC's 2013 report concludes that:

> Greenhouse gases contributed a global mean surface warming likely to be in the range of 0.5°C to 1.3°C over the period 1951–2010, with the contributions from other anthropogenic forcings, including the cooling effect of aerosols, likely to be in the range of −0.6°C to 0.1°C.

The IPCC Working Group II report of 2014 warned that 'No one on the planet will be untouched by the damaging effects of global warming in coming decades' as a result of threats to humans including flooding, storm surges, droughts and heat waves.[32]

Human-driven fossil fuel consumption and forest fires are the main causes of carbon dioxide build-up, with methane releases as a by-product of agriculture (predominantly from rice, cattle and sheep) also constituting significant sources. The resulting rise in atmospheric carbon dioxide levels is widely believed by atmospheric scientists to account for the fact that, on the basis of WMO data up to 2007, the decade from 1998 to 2007 was the warmest on record, with the global mean surface temperature for 2007 estimated at 0.41°C above the 1961–90 annual average of 14.0°C. Indeed, up to 2008, 12 of the warmest years on record had occurred in the preceding 14 years. While the geological record shows that climatic changes have taken place regularly, most notably during ice ages, modern changes are occurring at an unprecedented rate and appear to be linked to increasing levels of pollution in the atmosphere. The UN Environment Programme (UNEP) estimates that, by 2025, average world temperatures will have risen by 1.5°C above a 1990 datum. There are also implications for the disruption of important global sea currents, and the energy transfer and food production functions that they perform.

Extreme weather damage, the rising occurrence of which may be attributable to a changing climate, looks set to exert an increasing impact on grain output in China, the world's biggest grain producer. An analysis by Zheng Guoguang (head of the China Meteorological Administration), published in December 2009 in the magazine *Qiushi* ('Seeking Truth', published by the Central Committee of the Communist Party), found that climate-driven factors are already causing grain output to fluctuate by about 10% to 20% from longer-term averages. With global warming projected to intensify droughts, floods and pests, this is thought likely to widen to between 30% and 50%. This is just one of many examples across the world of empirical or theoretical studies addressing climate-driven instability in food production, studies that are near-unanimous in highlighting the potential for social and economic risks and/or hardships.

Rising temperatures lead to more extreme climatic events, including, for example, record heat waves, the melting of ice, rising sea levels, more destructive storms and increasingly intense rainfall events, which result in major flooding. Further consequences include respiratory problems in inner cities, loss of crop productivity and increasing risks to infrastructure and investments of all kinds. Evidence of the effects of warming and climate change are everywhere. The July 1995 heat wave in Chicago, when temperatures reached 38°C to 41°C on five consecutive days, claimed more than 500 lives and helped shrink the 1995 US corn harvest by some 15%, or $3 billion. There has been a recorded rise of about 1°C in the temperature of the world's oceans during the 1980s. Arctic ice was 6–7 metres thick in 1976 but had decreased to 4–5 metres by 1987. On 19 August 2000, *The New York Times* reported that an icebreaker cruise ship had reached the North Pole only to discover that this famous 'frozen' site was now open water.[33] The Arctic sea ice has thinned from nearly 2 metres thick in 1960 to scarcely 1 metre in 2001. Some scientists predict that, within 50 years, the Arctic Ocean could be ice-free during the summer. This, of course, has major implications for polar bears and other organisms and ecosystems that have evolved in close relationships with predictable patterns of ice formation and melt. In Europe's Alps, the shrinkage of the glacial volume by more than half since 1850 is expected to continue, with these ancient glaciers largely disappearing over the next half-century. Predictions about the loss of ice and snow on Tanzania's iconic Mount Kilimanjaro are being realised, with increasing implications for tourism. Meanwhile, retreating glaciers in the Himalayan range in central Asia are threatening to reduce the dry season flow of the River Ganges by as much as 50%, with serious consequences for the millions of people dependent upon it. As 'the world's water tower', glacier and snow loss across the wider Himalayas, exacerbated by poor upland land use decisions, may have serious consequences for major rivers and their billions of dependants across the wider Asian continent.

All of this extra water has to go somewhere, and it contributes substantially to the phenomenon of sea level rise. During the twentieth century, the sea level rose by 10 to 20 centimetres, more than half as much as it had risen during the preceding 2,000 years. Current sea level rise has occurred at a mean rate of 1.8 millimetres per year for the past century, and more recent estimates suggest an increasing rate of between 2.8 and 3.1 millimetres per year between 1993 and 2003. If current estimates are correct, sea level could rise by as much as 1 metre during the twenty-first century, with the coastline retreating, on average, by 1,500 metres and some smaller islands becoming uninhabitable. In central London, the Thames barrier, a set of ten movable gates constructed across the river in 1980 to prevent upriver tidal surge flows, has been closed with increasing frequency since its inauguration. Although the situation has been compounded by the downward tilt of the land mass of south-east England, the fact that the barrier was shut three times a year on average in the first ten years of operation but that this frequency has now doubled provides some indication of the potential consequences of rising sea levels exacerbated by increasing atmospheric energy contributing to storm surges.

A further consequence of higher temperatures is more energy driving storm systems. From 1920 to 1970, there were an estimated 40 major storms per year; between 1985 and 2000, the northern hemisphere experienced close to 80 storms a year, representing a doubling in less than a generation. The annual number of Atlantic hurricanes is higher now than at any time in the last 1,000 years.[34] With rising frequency has also come increased force and consequent economic damage, as was shown by the three powerful and damaging winter storms in France in December 1999 and by 1998's Hurricane Georges in Central America. Hurricane Katrina, making landfall primarily in the state of Louisiana and resulting in the inundation of the city of New Orleans in August 2005, was one of the deadliest natural disasters in US history, leaving 1,836 people dead and a further 705 missing, with effects felt more widely including an estimated US$1 billion to US$2 billion of damage in the neighbouring low-lying state of Florida. As a consequence of higher-energy storms and other phenomena relating to climate change, natural disasters are on the increase. For example, Munich Re, one of the world's largest reinsurance companies, reported that three times as many great natural catastrophes occurred during the 1990s than in the 1960s, economic losses increased eightfold, and insured losses multiplied 15 times. Munich Re's natural catastrophe figures for 2007 demonstrate higher losses despite the absence of mega-catastrophes, with cumulative events resulting in overall economic losses of US$75 billion, in line with the rising trend in natural catastrophes. Climate change is now recognised as a major risk contributing to rising claims for property damage worldwide.[35] We are seeing wider economic fall-out from climate change, but it is projected to get much worse as a consequence of environmental pressures already in place, unless there are serious interventions. At this rate of growth, by 2065 the amount of damage would exceed the projected gross world product; well before then, the world would face bankruptcy. Lester Brown, former president of the Earth Policy Institute in the US, suggested that 'Nature was levying a tax of its own on fossil fuel burning'.[36] The UK government's 2006 Stern Review,[37] discussed in more detail below, highlighted the economic benefits of early action to avert the worst consequences of climate change which would otherwise dwarf preventative costs.

An apparently small temperature rise in sea surface water of less than 1°C can lead to the effect of 'coral bleaching', the mass death of corals; bleaching has already devastated huge areas of reefs in the Caribbean, Indian and Pacific Oceans. If coral reefs continue to die, oceanic ecosystems will be altered, directly affecting the fisheries that depend on the coral reefs as nursery grounds. And, as fisheries fail in both tropical and temperate waters, we will need to travel further to exploit remaining grounds, in turn increasing the potential for conflict and driving the land ever harder to produce farmed crops to make up the shortfall in protein and other essential food groups, and all the while expending more energy, which feeds back in turn to climatic instability.

Beyond these physical, meteorological and thermal effects, climate change is also exerting a range of other indirect pressures on the Earth's ecosystems. Significant among these is ocean acidification, sometimes described as the 'evil twin' of climate change. Elevated carbon dioxide concentrations in the atmosphere are mirrored by levels of dissolved gases in the world's oceans, which are contributing to an ongoing decrease in their pH. Between 1751 (more or less coinciding with the onset of industrialisation) and 1994, mean global surface ocean pH is estimated to have decreased from approximately 8.179 to 8.104. This net change in pH of −0.075 may seem small, but it is equivalent to changes that occurred during the last 300 million years. Ocean acidification is responsible for a wide range of observed changed mineral flows and biological responses that have significant wider consequences for ecosystems, including, for example, making the process of shell formation increasingly difficult. If pH falls by a further 0.3 to 0.4 units by the end of the century, as is projected if current greenhouse gas emission trends are not mitigated, the consequences would be disastrous.

Among other diverse likely impacts of a changing climate, due largely to more energy in the atmosphere associated with increasingly severe storms, droughts and other extremes of weather, are massively enhanced erosion, including increasing rates of loss of productive soils, as well as eutrophication and habitat loss in aquatic ecosystems as a result of nutrients carried into them by sediment particles.

Less tangible problems

Of course, not all of the abuses of the airspace are chemical, but these other issues are far from trivial. Some tropospheric pollution, such as noise, may be physical.

It is not just gases and fine particulates that are conveyed by the connective medium of air, but also acoustic energy, from the hum of summertime bees and the song of birds to human-made music and the intrusive drones of motorways, drilling activities and aviation. Noise is one of the key drivers of nuisance complaints in neighbourhoods, but its ramifications may be far greater than inconvenient disturbance. Noise is increasingly being recognised for its impacts on wildlife and human health and wellbeing in addition to, or indeed indirectly as a consequence of, its nuisance value. The effects of noise pollution on wildlife are diverse, but they include a potential threat to health through chronic stress and hence to the long-term survival of animals in particular.[38] In ecosystems, it is suggested that all organisms have their own specific 'aural niche'; this includes not merely the noises that they make but also the acoustic environment to which they are adapted, so disturbance to this soundscape could be detrimental.[39] Certainly, studies on the effects of snowmobile noise on white-tailed deer in Minnesota revealed that, although the animals habituate to a certain extent, they increase their home range size and avoid active snowmobile trails.[40] Other observed adverse effects of noise on animals include damage to hearing, disorientation and alarm, reduced population densities of rodents, a decrease in the reproductive success of birds close to significant sources of noise, injury through panic reactions, avoidance of important feeding habitats, and possible disruption to animal communication.[41] Elaborating upon this literature is beyond the scope of this chapter, but clearly there is a case to be answered regarding the intrusion of society's noises on the natural world.

Noise has also been considered a pollutant in the human environment for decades. More than 30% of the European population experience noise levels that disrupt sleep or speech, mainly from road traffic, railways, residential areas, industry, aircraft, ports, construction sites, and sport and leisure facilities, with 20% of the population regularly exposed to noise levels that scientists consider an unacceptable health risk.[42] There are more serious health implications: for example, a study has found that exposure to aircraft noise at night for more than 20 years could increase the risk of heart disease and stroke, and risk also increased for those constantly exposed to road traffic, although air pollution may be a significant compounding factor in this outcome.[43]

There is also rapidly increasing interest and concern about the way in which modern society is illuminating the lower atmosphere; light pollution, also known as photopollution or luminous pollution, is receiving growing attention. Initially, artificial light was seen as a universally good thing, lighting up dark spaces and highways during the hours of darkness. However, humans and other organisms evolved with natural regimes of light and darkness. Departures from this regime are beginning to be implicated in a wide range of adverse impacts, leading to a definition of light pollution as 'Degradation of photic habitat by artificial light'.[44] Thinking of the natural regimes of darkness and light as part of an atmospheric 'photic habitat' refocuses attention on the basis and potential magnitude of impacts likely to arise from disruptions to it.

The disruption of darkness is known to result in diverse ecosystem perturbations, including the familiar sight of flying insects disoriented by street lamps and opportunistic predation by insectivores such as spiders. However, light pollution poses a serious threat to nocturnal wildlife. It can confuse animal navigation, alter competitive interactions, change predator–prey relations and contribute to physiological harm,[45] as well as having a far more profound and pervasive impact on ecological dynamics through the alteration of the natural diurnal patterns of light and dark.[46] For example, night lighting can disorient moths sufficiently to seriously affect the reproduction of plant species with night-blooming flowers, leading to their progressive decline and to wholesale long-term changes to the ecology of illuminated areas.[47] Likewise, lights on tall buildings or other structures are known to disorient migrating birds, potentially contributing to their mortality.[48] Night lighting also affects the ability of sea turtle hatchlings to find their way to the sea,[49] along with a wide range of other observed adverse ecological impacts. There is also growing evidence that nature is becoming confused by artificial light sources, adapted as it is to particular spectra of light as well as to polarisation.[50]

Humans, of course, also evolved within and remain integral elements of the atmosphere and wider biospheric system. So it should not be surprising to find that the relatively recent shifts that we have made in the natural rhythms of dark and light have implications for our health and wellbeing. Light pollution, in terms of both excessive or prolonged light exposure and changes in the spectrum of light to which we are exposed, is implicated in effects as varied as increased incidences of headaches, fatigue among workers, medically defined stress, decreased sexual functioning and raised anxiety.[51] Indeed, in 2007, the World Health Organization's International Agency for Research on Cancer listed 'shift work that involves circadian disruption' as a probable carcinogen,[52] with multiple studies documenting a correlation between night-shift work and the increased incidence of breast, prostate, colon and lung cancer.[53]

There are also, of course, implications for energy usage from different lighting regimes within and outside buildings, and these in turn may significantly affect other facets of the atmosphere, including emissions of climate-active and other potentially problematic substances. So a reconsideration of lighting may have multiple implications for overall atmospheric pollution and net impacts on ecosystems and people. Furthermore, the concept of 'light trespass' addresses unwanted light entering a person's property, for example when bright sources shine across a neighbour's fence, blocking views of the sky and stars as well as contributing to sleep deprivation.

Visual blight, too, could also be considered a form of pollution mediated by the transparency of the troposphere, as well as the blocking out of dark skies and disorientation of organisms that navigate by dark and light. We will touch on these issues later in this book, though not in any greater detail at this stage.

The ways in which we abuse the atmosphere are therefore many and varied, and often below the 'radar' of common appreciation and control mechanisms.

Emerging challenges

As if the impacts of anthropogenic chemical, physical and biological emissions and other abuses of the air were not already diverse and complex enough, we are also becoming increasingly aware of new sources of problems.

Air is the ultimate communicator and distributor. Types of 'pollution' released into it, disseminating to neighbouring people and ecosystems, though regarded largely as a matter of nuisance, include dust, noise, light and visual blight. Recalling that historic concerns about pollutant gases were largely perceived as aesthetic, but with some association between bad smells and disease in the 'mal air', it is worth rethinking the ways in which we perhaps too readily dismiss these wider factors.

Dust, for example, comprises a diversity of particles fine enough to be whipped into the winds and carried for unknown distances. Some are natural, such as pollen, and some inorganic matter contributes to land-forming processes. Other constituents of the dust result from excessive wind erosion of poorly managed soils, not only polluting the air but contributing to respiratory problems, and fouling and clogging urban infrastructure. Also, as observed in the case of the Aral Sea, wind-blown dust bears with it a potentially toxic cargo of attached pollutants. Furthermore, poor agricultural land management practices enable the wind to bear away a key natural resource, contributing to massive soil loss on a global scale that threatens food security and overall societal resilience. Industrial, transport, domestic and other anthropogenic emissions are further sources of dust of often indeterminate but potentially hazardous content. And activities such as quarrying are known major dust sources, potentially blighting health and adjacent property values.

To meet more stringent requirements to treat waste streams and to avert inputs to landfill, there have been substantial increases in the composting of putrescible waste. Not only does composting reduce flows of organic waste into landfill, thereby reducing the generation of climate-active gases such as methane, it also helps recycle the organic and nutrient content of the waste materials into the useful product of compost. There are various techniques for industrial-scale composting, the most popular agricultural version of which is 'windrow composting'. This entails piling organic matter or biodegradable waste, including animal manure and crop residues, into long rows known as windrows. The windrows are turned regularly to mix and aerate them, improving their porosity and the drainage of moisture, as well as evening the temperature of the warm compost. However, notwithstanding the beneficial recycling of waste, there are growing health concerns about the implications for workers and neighbours of airborne microorganisms released from industrial-scale composting processes.

The dominant airborne particles of concern are bioaerosols that comprise particles of plant or animal origin as well as microbial content including living and dead bacteria, fungi, viruses, allergens, bacterial endotoxins, antigens and other toxins, pollen and various types of fibre.[54] Frequently, microorganisms are absorbed in dust particles and are transported along with them. Exposure to bioaerosols varies over time, although, unsurprisingly, studies demonstrate that aerial concentrations tend to be elevated downwind of outdoor composting facilities. Principal health effects of concern include respiratory symptoms, mucosal membrane irritation, skin diseases and immune responses, which are found to be associated with the distance residents live from outdoor composting sites. Emissions from composting can be minimised by reducing the agitation of composting piles as well as by watering to minimise dust. Further concerns result from the odours emitted from composting facilities, but these can be minimised by good oxidation, typically by including more fibrous matter. Facilities at composting sites close to where people live may also need to be enclosed in order to control emissions to the air.

Once one starts to look, there is a never-ending cavalcade of currently overlooked issues that may be of concern, including, for example, unintended negative impacts for flying organisms (principally migrating birds and bats) that arise from wind turbines erected as a contributory measure to protect the atmospheric system from greenhouse gas emissions.

The airspace that we share not merely with our fellow human beings but with all life is a key medium of communication and connection, profoundly influencing – and in turn shaped by – the actions of all that share it. Just as 'No man is an island'[55] within human societies, so no unit of the biosphere is discrete, no symptom of change stands alone, and feedback between them is an integral feature of the complex world of which we are a part. If we degrade habitat quality and extent, we also degrade the capacity of associated ecosystems to purify air and water, produce food, support diverse ecosystems, and provide economic and aesthetic potential. Through feedback loops in the biosphere, there are intimate links between climate change, ozone depletion, bioaccumulation of pollutants, depletion of natural resources and the sustenance of biodiversity. And these, in turn, impinge directly upon the quality of the air we breathe and use for so many industrial and other purposes, the availability or scarcity of food, fertility of croplands, land use potential, population pressures, education, disease transmission, population growth and healthcare.

By degrading the environment, not merely the air itself but the terrestrial and aquatic ecosystems that profoundly shape it, we degrade human potential. In turn, this places ever greater stresses on the environment. This cycle of causality, and the need to innovate to escape it, is a significant feature distinguishing sustainable development from basic 'environmental protection' measures predicated simply on minimising impact.

Into the Anthropocene

The cumulative impacts of human activities are beginning to have a dominant effect on the ecosystems and cycles of Planet Earth. This has led to many scientists now conceding that we are in transition from the Holocene, the geological epoch of the past 12,000 or so years since the end of the Pleistocene epoch, to the Anthropocene. The term 'Anthropocene' is an informal geological chronological term, coined originally by the ecologist Eugene Stoermer and popularised by the atmospheric chemist Paul Crutzen,[56] recognising that the influence of humanity on the Earth's atmosphere and ecosystems in recent centuries has been on a scale significant enough to constitute a new geological epoch.

This was not the first time that the scale of human impact was considered a sufficiently dominant influence: the term 'Anthropozoic Era' was used by the Italian geologist Antonio Stoppani in 1873,[57] and the term 'Homogenocene' used by Michael Samways in 1999.[58] Although the Anthropocene – the age of humans – has no clear start date, atmospheric evidence is strong enough to mark its commencement as the beginning of the European Industrial Revolution, although some scientists consider that it started with the rise of agriculture, and hence the widespread conversion of landscapes, in the Neolithic period.

Human influence over the characteristics of the global environment is profound – whether that influence is measured by the pace of conversion of the face of the Earth, the loss of habitats and species, the warming of the atmosphere and oceans and the contamination of all environmental media, including by substances new to nature, or judged by changes in atmospheric gas composition. We may be the authors of a new era. However, history will judge whether our impacts are framed purely by degradation through continued abuse, or else by an age of increasingly responsible and far-sighted stewardship.

5 | Managing our impacts on air

Global society has responded in a diversity of ways to rising pollution and other pressures on the atmosphere. These range from local nuisance through to mobilisation to address global threats, spanning issues of urban design, industrial process controls and land use.

A slow start

Concerns about human impacts on the air are far from new, extending back to around 2,000 years ago, when the Roman senate introduced a law according to which *aerem corrumpere non licet*: 'polluting air is not allowed'. The Roman poet Horace (Quintus Horatius Flaccus, 65 BC–8 BC) recorded the blackening of buildings due to smoke in Rome in classical times.[1]

In 1273, the burning of sea coal was prohibited in London due to its prejudicial implications for public health. There is some dispute about what this 'sea coal' actually was, as the term has been ascribed to coal washed up on beaches and thought to originate from exposed coastal coal beds, or it may relate to coal imported by sea, or it could be a term distinguishing coal from charcoal. However, the principle of averting noxious emissions from the substance in an urban setting remains the same. In 1306, King Edward I of England reinforced this prohibition through a royal proclamation banning the burning of sea coal in craftsmen's furnaces owing to the foul-smelling fumes produced.[2]

Beyond these two early examples, controls on air pollution are not perhaps as deep-rooted in history as the protection of water quality; the impacts on human health of water quality were recognised as early as the Middle Ages, resulting in the enactment of protective legislation. By contrast, air pollution was seen initially as being more a nuisance than a threat to human health. So, for example, in around 1590, Elizabeth I was 'greatly grieved and annoyed' by coal smoke in Westminster Palace, and banned the burning of coal in London while parliament was in session, again for largely aesthetic reasons.[3]

In both England and the US, a range of smoke control ordinances were enacted in the late 1800s and early 1900s. These laws comprised some of the first uniform statutes enacted for the control of air pollution, from which modern air pollution control programmes ultimately derive.

Responding to growing concerns about local air pollution

Legislative responses to ever worsening air pollution across the industrial and industrialising worlds only began to gain momentum with the widespread realisation that pollution represented not merely a nuisance but a health threat of crisis proportions. This was initially recognised by scientists and health professionals, but spread to wider publics and ultimately politicians. A range of prominent global air pollution episodes resulting from the build-up of pollutants was influential in the development of air pollution programmes. These included an acute air pollution event in 1930 in the Meuse River Valley, Belgium, in which high atmospheric concentrations of sulphur dioxide during a temperature inversion resulted in the deaths of 63 people, with thousands more falling sick. In 1931, the Manchester and Salford smog in Britain's industrial heartland led Charles Gandy, a Manchester barrister and chairman of the National Smoke Abatement Society (formed by the merger of the Coal Smoke Abatement Society and the Smoke Abatement League), to propose the concept of smokeless zones: designated zones particularly susceptible to pollution from combustion in which only smokeless fuels are permitted.

Air pollution received greater attention throughout the 1940s in the United States as smog became more evident in Los Angeles, reducing visibility, stinging eyes, and inducing respiratory discomfort, nausea and vomiting. In Donora, Pennsylvania, in 1948, high concentrations of sulphur dioxide coupled with a temperature inversion and foggy weather contributed to the deaths of 20 people due to cardiac and respiratory diseases, with around half of the town's 12,000 residents complaining of coughs, respiratory tract irritation, chest pain, headaches, nausea and vomiting. In Poza Rica, Mexico, in 1950, a natural gas plant inadvertently released hydrogen sulphide; this coincided with a temperature inversion and foggy weather, and killed 22 people with a further 320 hospitalised. And then, in 1952, the famous Great Smog of London formed during a five-day temperature inversion, during which the build-up of harmful acidic atmospheric aerosols resulted in 4,000 people succumbing to bronchitis, pneumonia, and respiratory and cardiac diseases.

These concerns cumulatively led to the creation in the US of some of the world's most comprehensive air pollution control laws. Responding to the gathering California smog, the State of California passed the first state-level air pollution law in 1947. In 1949, the issue received US-wide attention under the first National Air Pollution Symposium. This contributed to the US federal government instigating efforts to control air pollution, including mandating research programmes to explore the health and welfare effects of air pollution and passing the Air Pollution Control Act of 1955. This was superseded by the Clean Air Act 1963, which required the development of air quality criteria based on scientific studies and provided grants to state and local air pollution control agencies. The Motor Vehicle Air Pollution Control Act 1965 required the establishment of automobile emission standards, and, in 1967, the Federal Air Quality Act established a framework for defining 'air quality control regions' based on meteorological and topographical factors contributing to localised air pollution. Then, in 1970, the Environmental Protection Agency (EPA) was established by executive order under Richard Nixon's presidential administration, marking a dramatic change in national policy regarding the control of air pollution, which now included the stringent enforcement of air pollution laws and required damage already inflicted on the natural environment to be repaired. Passage in 1970 of the Clean Air Act Amendments marked the beginning of modern efforts to control air pollution.

In the UK, there is a particularly long legacy of statutory protection of the air and the benefits that it confers upon people, including, for example, a thirteenth-century royal proclamation relating to the 'excessive burning of sea coal' and various legal instruments instituted since the nineteenth century.[4] Action on air pollution began to take effect from the 1860s with the Alkali Act 1863, and its successor Alkali etc Works Act, 2nd Act 1874, reflecting growing concerns about the impacts of unconstrained industrialisation on urban air quality (particularly of smoke and sulphur dioxide emissions) and their effects on public health. This was joined by the Public Health (Smoke Abatement) Act 1926. And, as noted earlier, the reaction to the Manchester and Salford smog of 1931 led to the concept of the smokeless zone. In the UK, these are now known as 'smoke control areas', in which it is permissible to burn only authorised fuels such as anthracite, semi-anthracite, gas and low volatile steam coals, as well as oil or other liquid fuels in specially designed fireplaces.[5] Enactment of the UK's Clean Air Act in 1956[6] occurred largely in response to London's Great Smog event of 1952. A growing body of legislative responses in the UK during the 1990s included the Environmental Protection Act 1990 and the Environment Act 1995, with public concerns generating a growing need for air quality standards to manage risks associated with emerging technologies.[7] The Pollution Prevention and Control (England and Wales) Regulations 2000,[8] implementing in part the European Union (EU) Integrated Pollution Prevention and Control Directive (1996/1/EC), require an integrated approach not to environmental media but to industrial processes affecting them, with the aim of preventing pollution risks being transferred unthinkingly from one medium to another without regard for cross-media optimisation.

A similar trajectory of legislative responses to air quality concerns can be observed throughout the developed world. For example, Japan's body of air pollution control law collectively aims to protect the nation's health and to conserve the human environment. Examples of measures put in place in Japan include regulations for smoke, soot and dust emissions from factories and business establishments, implementation of countermeasures to address hazardous air pollutants, and stipulation of allowable limits for automobile exhaust gases, as well as providing liability in damages to protect victims of air pollution.[9] In Australia, the National Environment Protection Council, set up in 1994 and comprising ministers representing participating commonwealth, state or territory governments, established the country's first uniform national ambient air quality standards for the six key air pollutants to which most Australians are exposed: carbon monoxide, ozone, sulphur dioxide, nitrogen dioxide, lead and particulates.[10]

Various other air pollution acts around the world include the Canadian Clean Air Act 1970, the US Clean Air Act 1963, New Zealand's Clean Air Act 1972, the EU Air Quality Management framework, and their many subsequent amendments. Much of the rationale behind EU directives of this era had been to bring about consistency in environmental practice to prevent member states gaining trading or economic advantages over others by avoiding the costs of controlling air and other pollution; however, one of the net results was further incremental progress towards addressing broader facets of the air and the wider atmospheric system. Air quality management frameworks are being progressively established in developing nations such as India and South Africa, where bad urban air quality contributes substantially to poor public health and premature deaths.

Acting across national borders

As the scale of harm stemming from a range of atmospheric pollution issues became increasingly apparent, so too did the need for greater collaboration across national and continental boundaries to bring about concerted change. Legislative responses to atmospheric pollution therefore began to extend beyond national borders.

As one example, concerns grew internationally about an elevation in radioactive contamination of the atmosphere as it became clear that atmospheric nuclear weapon tests had almost doubled the concentration of carbon-14 in the northern hemisphere as well as contaminating other environmental media. This led to international action that culminated in the enactment of the Limited Test Ban Treaty (LTBT), signed in Moscow in July 1963 and ratified during autumn 1963 by the Soviet Union, UK and US. The LTBT prohibits all test detonations of nuclear weapons except underground (i.e. banning tests in the atmosphere, in outer space and underwater). Also known as the Comprehensive Test Ban Treaty, the LTBT served to slow down the dangerous pace of the arms race and opened the way to disarmament, as well as to limiting contamination of the air and other environmental media with radioactive waste. Most countries have since signed and ratified the treaty, though nations known to have tested nuclear weapons but that have not signed the LTBT include France, China and North Korea.

Rising awareness and concerns about transboundary acid rain issues were prominent in driving international action in the 1970s. Particularly influential examples included transboundary acid pollution arising from emissions from the UK implicated in forest damage in Norway and Sweden, and reciprocal damage between Canada and the USA. Aerial plumes were also observed to have an impact within national boundaries, for example the fallout from emissions in the acid-vulnerable uplands of Wales resulting from industrial and urban emissions downwind in the valleys of south Wales, exacerbated by the planting of coniferous trees in vulnerable areas. Unilateral national action alone was insufficient either to avert the damage or to satisfy aggrieved forest interests, affected communities and their political allies, leading to development of the United Nations Economic Commission for Europe (UNECE) Convention on Long Range Transboundary Air Pollution, adopted in 1979 and coming into force in 1983.[11]

Discovery of the Antarctic 'ozone hole' in 1985, and the need to fix this 'hole in the roof of the world', was a particular cause of panic. This reinforced the need to prioritise mobilisation of the international community to address a global problem for which only a worldwide cap on emissions of ozone-depleting substances would suffice. It is ironic that modern lifestyles can exacerbate ground-level ozone concentrations, raising a range of connected health, aesthetic and crop production problems, while simultaneously contributing to the thinning of naturally occurring ozone in the lower stratosphere, which then results in greater incidence of damaging radiation reaching the planet's surface. As we have seen when addressing *Abuses of the air* (Chapter 4), this ozone thinning is an inadvertent consequence of the incautious release of long-lived synthetic chemicals that catalyse the breakdown of ozone when circulated into upper layers of the atmosphere.

Responding unilaterally to a 1976 report by the United States National Academy of Sciences, which had marshalled a range of scientific evidence effectively proving the mechanisms underlying stratospheric ozone depletion,[12] a number of countries agreed to lead on a ban on the use of chlorofluorocarbons (CFCs) in aerosol spray cans.[13] Measures taken by these countries, including influential nations such as the USA, Canada, Sweden, Denmark and Norway, paved the way for more comprehensive global regulation – in opposition to powerful lobbying by the halocarbon industry, particularly in the USA and Germany – and were ultimately driven by increasingly dire revelations about the magnitude of harm already inflicted on the ozone layer. Though imperfect, the ensuing 'Montreal Protocol' – the Montreal Protocol on Substances that Deplete the Ozone Layer,[14] adopted in 1987 and coming into force in 1989 – was effective in forming a global consensus about phasing out the production and use of CFCs, halons and other ozone-depleting chemicals, including carbon tetrachloride and trichloroethane. Kofi Annan, former Secretary General of the UN, stated that 'perhaps the single most successful international agreement to date has been the Montreal Protocol'.[15] Treaties stemming from the Montreal Protocol have since become the most widely ratified in UN history. They have resulted in a steady decline in ozone-depleting chemicals after their peak levels in 1994 and, it is believed, will allow the recovery of the ozone layer by 2050[16] or 2068[17] if their conditions continue to be observed.

International action also ensued to address a range of other issues of mounting concern. For example, across the EU, the Dangerous Substances Directive (67/548/EEC) was formulated following rising disquiet about emissions of problematic chemicals to all environmental media. Further international agreements and action resulted from concerns about the dispersal of radioactive fallout from the 1986 Chernobyl explosion and about continuing issues relating to sulphur and nitrogen emissions to air (Sulphur Dioxide and Suspended Particulates EU Directive 80/779/EEC) from large combustion plants (Large Combustion Plant Directive 88/609/EEC). In the 1990s, emerging fears about volatile organic compounds (VOCs), oxides of nitrogen (NO_x), PM_{10} and other fine particulates emitted from vehicles renewed focus on ambient concentrations of air pollution, including ground-level ozone, and led to rising concerns about climate change.

Continuing transnational legislative responses in the 1990s included the European Commission (EC) Framework Directive on Ambient Air Quality Assessment and Management 96/62/EC.[18] Associated with this is a series of 'daughter directives' that serve to control levels of specific pollutants and to establish monitoring procedures, methods of measurement and calibration in order to achieve comparable measurements across EU member states. The framework directive and its daughter directives were subsequently merged into a single Ambient Air Quality Directive.[19]

Of course, these international commitments need to be transposed into national legislation and strategies to ensure consistency and integration across levels of international, national, regional and local government and enforcement. For example, the air quality management framework in the UK implements a range of EU directives, significantly including the framework directive. The UK's air quality management framework evolved from its Environment Act 1995, establishing a process for identifying and remediating areas of poor air quality and requiring the publication of a National Air Quality Strategy (NAQS).[20] The first NAQS for England, Scotland, Wales and Northern Ireland, *The United Kingdom National Air Quality Strategy*, was published in March 1997 and set out a range of requirements for the Environment Agency, local authorities and the devolved administrations (Wales, Scotland and Northern Ireland).[21] Government commitments for implementing these requirements include the use of sound science and of the precautionary, 'polluter pays' and sustainability principles. The main focus of the 1997 NAQS was the setting of health-based standards for eight pollutants;[22] the quality standards for these were determined by a government-commissioned Expert Panel on Air Quality Standards (EPAQS). The Air Quality Regulations 1997[23] provided the legal framework. Amendments and updates are made to the strategy from 'time to time': a second edition of the NAQS was published in 2000, bringing into effect the new UK-wide air quality objectives for the eight specified pollutants. An additional revision was published in 2007,[24] and updated policy and technical guidance was published in 2009.[25] Implementation requires action at the national level, with central government and its agencies establishing protocols and standards. Action is also required at the local scale through the collaboration of local authorities, the

Environment Agency and other stakeholders such as the Department for Transport, which all have responsibilities for specific local action. Consequently, local authorities are required periodically to review and assess a prescribed set of local air pollutants predominantly arising from transport, industry and domestic sources through a statutory Local Air Quality Management (LAQM) process.

A climate of change

Of course, the 'granddaddy' of all international mobilisations occurred to address the rising, planet-wide and long-term threat of climate change. Climate change is a truly pan-global threat, driven not merely by a greater diversity of chemicals entering the atmosphere but also by a host of societal activities ranging from combustion for energy generation, industry, home heating and transport to changes in land use, clearance of forest and peatlands, and changes in floodplain hydrology. However, compared with societal responses to local air quality problems and ozone depletion, management responses to the emerging threat of climate change are at a far more preliminary and fragmented stage of evolution.

Recognising the need to integrate knowledge and coordinate resultant recommendations on a fully international basis, the Intergovernmental Panel on Climate Change (IPCC)[26] was set up in 1988 at the request of member governments. The IPCC was established as a scientific intergovernmental body by two organisations of the UN – the World Meteorological Organization (WMO)[27] and the United Nations Environment Programme (UNEP)[28] – and subsequently endorsed under UN General Assembly Resolution 43/53. The purpose of the IPCC is to provide comprehensive assessments of current scientific, technical and socio-economic information worldwide about the risks of climate change caused by human activity, its potential environmental and socio-economic consequences, and possible options for adapting to these consequences or mitigating the effects.[29] The IPCC does not conduct its own original research or monitoring, but literally thousands of scientists and other experts from more than 120 countries contribute voluntarily to its work.[30]

The principal forum for international agreement on tackling climate change is the United Nations Framework Convention on Climate Change (UNFCCC). Created in 1992, the UNFCCC is a forum for negotiation on four key areas: 1) adapting to climate change; 2) finance to enable action on mitigation and adaptation; 3) mitigation (i.e. reduction) of greenhouse gas emissions; and 4) technology development and transfer to allow green development. As of November 2013, 195 countries had joined this international treaty. Key outputs of the IPCC include a series of special reports on topics relevant to the implementation of the UNFCCC, such as risks of extreme events and disasters, renewable energy sources and climate change mitigation. To date, the IPCC has published five comprehensive peer-reviewed assessment reports (in reality, multi-volume reports) reviewing the latest climate science, in addition to a number of special reports on particular topics. The first IPCC assessment report was published in 1990; this was followed by a supplementary report in 1992, a second assessment report in 1995, a third in 2001, a fourth in 2007, and a fifth in 2014 – this last report increased the degree of certainty that human activities are driving the warming of the world from 'very likely' (or 90% confidence) in 2007 to 'extremely likely' (or 95% confidence).

In light of rapidly growing awareness of the scale of threat of anthropogenic climate change to humanity, the control of emissions of greenhouse gases was a significant theme debated at the 'Earth Summit': the United Nations Conference on Environment and Development (UNCED) held in Rio de Janeiro in 1992. At the Earth Summit, it was agreed that, by 2000, countries would stabilise carbon dioxide emissions at 1990 levels. All climate change exceeding natural variability is, of course, a matter of concern, and there was international consensus about the need to restrain emissions so that global temperature rise does not exceed 2°C in order to reduce the chances of runaway climate change due to positive feedback mechanisms in the environment. The potential for so-called rapid climate change is a particular concern, as largely unpredictable tipping points are likely to be reached in linked phenomena such as ocean circulation in the North Atlantic and El Niño and La Niña episodes in the south Pacific, which would have a profound influence on climate. This is, however, a game of high stakes and even higher uncertainties.

This agreement to limit the extent of global warming took place under the aegis of the UNFCCC, with subsequent agreement under the 1997 Kyoto Protocol (an international agreement linked to the UNFCCC and signed in Kyoto in December 1997, coming into force in February 2005) to binding targets to reduce greenhouse gas emissions. The Kyoto agreement recognises that the burden of reducing greenhouse gas emissions falls most heavily on industrialised countries, taking into account their responsibility for past emissions, and that developing nations should have a right to increase their emissions to 'catch up' on development. Under the 1997 Kyoto Protocol, 37 industrialised countries formally agreed to control emissions of climate-active gases in order to address an overall 5% global reduction below 1990 levels for the first commitment period (2008–12). The Kyoto Protocol has thus far been ratified by 184 parties of the convention. The USA did not ratify the treaty, although a number of US states have made independent commitments to strive to meet the required reductions signalled by the protocol.

However, strategically important though they are for climatic stability and averting the worst effects of climate change, reductions alone are insufficient to address the threats. Climate change is happening, and it is felt most acutely by low-lying, arid and other countries and regions with natural vulnerability to rising sea levels, acidifying oceans and shifting weather patterns. Adaptation to climate change is another essential response. The UNFCCC commits all of its signatories to formulate, implement, publish and update adaptation measures, as well as to cooperate on adaptation, including putting in place a variety of support mechanisms for the implementation of adaptation measures in developing countries. The Cancun Adaptation Framework, adopted in 2010, agreed that adaptation should be given the same attention as mitigation, prioritising the reduction of vulnerability and the increase of resilience in developing countries.

In December 2009, representatives of 193 countries, including 130 heads of state, gathered in Copenhagen for the UN Climate Change Conference, known as COP15. The purpose of COP15, attended by some 40,000 formal participants from across the world, was to advance ongoing and evolving commitments to curbing global emissions of climate-active gases in the shape of new national targets. However, this mooted 'last best hope' for an agreement that would put collective global wellbeing ahead of the self-interest of individual nations began to unravel in the face of posturing from some industrialised nations and mistrust between the developed and the developing world. Nevertheless, developed countries pledged to provide climate aid to developing countries. Also, a weakened, non-binding Copenhagen Accord emerged, emphasising the strong political will to combat climate change urgently in accordance with the principle of common but differentiated national responsibilities. This included a comprehensive adaptation programme, including international support, to help reduce vulnerability and build resilience in the most climate-vulnerable developing countries. There was also strong consensus, especially among industrialised nations, about the need to hold global temperature rise to no more than 2°C. US president Barack Obama noted that the most important result in Copenhagen overall was that large emerging economies began, 'for the first time', to open up to taking responsibility for limiting their growth of greenhouse gases. Obama's speech seemed to allude mainly to China, which today is the world's largest emitter of greenhouse gases (although, of course, the US itself had failed to ratify the Kyoto Protocol a dozen years before). Although sub-optimal, with actions from the global meeting disproportionate to the rhetoric of crisis, COP15 was at least a step forwards in terms of tackling climate change on a concerted global basis.

Beneath the consensus, intense debate, disagreements and negotiating positions remain. China, for example, is leading appeals for rich countries to live up to their commitments on climate change. As one of the 195 countries gathering under the UNFCCC in Warsaw in November 2013 – a meeting imbued with a renewed sense of urgency in the wake of Typhoon Haiyan, which killed more than 6,000 people in the Philippines – China's delegation reiterated its pledge to play a more active role in addressing climate change, urging developed countries to honour prior commitments to finance climate change initiatives and technology transfer as a precondition of any further negotiations.[31] This strong negotiating position on behalf of developing nations is not without foundation, as, in 2009, the developed nations had promised to spend US$100 billion per year until 2020 to help developing countries adapt to climate change; however, the Green Climate Fund, which was set up to channel this money, remains underdeveloped and under-resourced. Fewer countries have remained signatories of the Kyoto Protocol for the second commitment period (2013–20), yet UNFCCC is continuing to develop a new protocol with agreed outcomes for all parties by 2015.

By this time there was already consensus that global emissions would probably need to be cut by 60% relative to a 1990 baseline, the UK government announcing a commitment in March 2007[32] that it would cut emissions by 60% by 2050 in response to a recommendation from its Royal Commission on Environmental Pollution (RCEP) in 2000.[33] This commitment has since been upgraded to 80% in the light of subsequent advice.[34] (Meanwhile, the RCEP itself was axed by the self-styled 'greenest government ever' in 2010.) The situation is complex, as developing nations need, and have a moral right, to ramp up energy use to meet their growing demands and aspirations, in addition to wishing to trade in resources such as forests that are major global carbon sinks. Meanwhile, the developing world has to face up to the fact that its historical pathway of development is responsible for the bulk of anthropogenic climate disruption, the worst impacts of which are likely to be borne by developing countries. Consequently, some authorities are calling for a 90% cut in carbon dioxide equivalent emissions by developed countries to allow the developing world to 'catch up' in economic and technological terms.

Of course, control of climate-active emissions depends not merely on abating emissions from societal activities but also on retaining carbon where it is sequestered in the lithosphere and biosphere. Forests and wetlands are particularly significant sinks of carbon, and many of the densest and as yet economically unexploited deposits are laid down in developing countries. There are moral as well as legal issues associated with developed nations, much of whose wealth is founded on a climate-destructive pathway of development, demanding that developing nations curb their aspirations for economic emancipation. So novel and economically tenable schemes are required to create incentives for developing nations to retain carbon deposits in situ. It is for this reason that the UN instigated its REDD+ mechanism – the Collaborative Programme on Reducing Emissions from Deforestation and Forest Degradation in Developing Countries.[35] REDD+ recognises that land use and biocapacity are key to the global carbon cycle, and that many solutions to climate change will come from carefully managing our use of ecosystem services.

Deforestation and forest degradation account for nearly 20% of global greenhouse gas emissions, driven by agricultural expansion, conversion to pastureland, infrastructure development, destructive logging, fires and other causes. This exceeds emissions from the entire global transportation sector, and is second only to emissions from the energy sector. Therefore, climate stabilisation is practically impossible without reducing emissions from the forest sector, in addition to implementing other mitigation actions. REDD was established as a mechanism to create financial value for the carbon stored in forests, offering incentives for developing countries to reduce emissions from forested lands and to invest in low-carbon paths to sustainable development. This was superseded by REDD+, which goes beyond deforestation and forest degradation to include conservation, sustainable management of forests and enhancement of forest carbon stocks. REDD+ includes a wide range of mechanisms, including the Forest Carbon Partnership Facility (FCPF) and the Forest Investment Program (FIP), which is hosted by the World Bank. These in turn could open up market mechanisms through which payments made by industrialised nations for emission offsets would reward developing countries for protecting ecosystems that are of value for many purposes, including the safeguarding of wildlife and traditional livelihoods, as well as sequestration of globally significant quantities of carbon. Developing countries enjoy many co-benefits as a result of preserving their forests under REDD+, including a range of ecosystem services such as water regulation, soil protection, non-timber forest products (food and fibre, for example), climate regulation and biodiversity, all of which provide revenue and sustain livelihoods.

But, of course, big set-piece events are not the sum total of global agreements. COP15 acknowledged that many local initiatives are already moving ahead without global agreements. These include programmes and commitments undertaken by cities, proactive businesses and geopolitical regions. The United Arab Emirates' Masdar City[36] provides a prominent example of a municipal and national commitment to achieving carbon neutrality in urban design.

The geographical scale, scientific breadth and political inclusivity of the IPCC have been heroic, informing the often hesitant progress being made in tackling the challenges posed by climate change. Indeed, the IPCC was joint winner of the Nobel Peace Prize in December 2013, shared with former US vice president Al Gore for his contribution to public awareness of the issue. IPCC reports are not without their critics, however; among other issues, there has been criticism of their inference of conclusions based on often sparse data and of flaws in some predictions (most notably the projected date of melting of Himalayan glaciers). This is to be expected within the process of scientific progress and the posing and testing of hypotheses, particularly as they relate to promoting action rather than reinforcing timidity in political responses. Indeed, former IPCC chairman Professor Bob Watson stated that:

> The mistakes all appear to have gone in the direction of making it seem like climate change is more serious by overstating the impact. That is worrying. The IPCC needs to look at this trend in the errors and ask why it happened.[37]

Conversely, other critics express concern that some IPCC reports tend to underestimate dangers and understate risks as they are founded only on 'lowest common denominator' findings.[38]

Consensus and measures to address climate change become decidedly patchier when it comes to national implementation, at least in developed countries where there are heavily vested interests in preserving an energy-intensive status quo. For example, the UK will meet its 2020 emissions reductions agreed under the second commitment period of the Kyoto Protocol, but largely thanks to a 'dash for gas': a switch in energy generation towards natural gas during the 1990s following privatisation of the electricity generating industry. This 'dash for gas' was driven by: regulatory changes allowing gas to be used as a fuel for power generation; economic considerations relating to the speed at which gas turbine power plants could be built compared with coal and nuclear power stations; innovation of combined-cycle gas turbine generators that provided higher relative efficiencies at lower capital costs compared with other conventional generating technologies; and the relatively recent development of North Sea gas. So the UK's contribution to Kyoto Protocol targets is almost entirely coincidental rather than deliberate. Meanwhile, because of increasing consumerism, the 'carbon footprint' that the UK and other developed economies, as major importers, imprints upon producing countries does not seem to have diminished at all. So any self-congratulation by developed world economies rings hollow: real systemic commitment to change is not yet evident. Indeed, under the austerity drive of the early 2010s, commitments such as environmental and social taxes on the UK energy generation sector have been the subject of government U-turns that once again posit short-term profit making over and above any serious long-term obligation. In fact, many purportedly 'green' taxes fail the simple test of hypothecation, whereby environmental 'bads' are taxed as a disincentive and revenues are recirculated as subsidies for environmental 'goods'. The UK government has even cut massively the leading 'feed-in tariff', which subsidises the installation of renewable generating technologies such as home- and commercial-scale solar panels on the back of a general tariff on other forms of energy generation – one of many reversals of commitments made by the Conservative–Liberal Democrat coalition's 'greenest government ever'.

Meanwhile, investment in, consent for and exploitation of cheap shale gas through the controversial 'fracking' (hydraulic fracturing) process proceeds with strong government support in the US and much of Europe. This is an illogical response to stated commitments to move away from a reliance on fossil carbon-based energy, especially as there are substantial unknowns about the process's implications for wildlife, geological stability, the eventual fate of large volumes of water and associated additive chemicals, and adjacent property values. A further downside to the release of all this shale gas is that excess US coal is now exported worldwide – about half of it to Europe and a quarter to Asia – resulting in 'dirtier' emissions in those places affecting the global climate, and in turn affecting global populations now and into the future.[39]

Meanwhile, continued dependence on carbon-intensive emissions across Europe, the US and other nations is being perpetuated by investment in carbon capture and storage technologies, stripping carbon dioxide from emissions and injecting it into deep seams under the sea bed, the long-term and wider impacts largely unknown. So the signals are not merely mixed and fragmented, but many directly contradict common knowledge and stated commitments about addressing climate change as a strategic priority.

The California Global Warming Solutions Act of 2006 also stands as a landmark, at odds with the lack of federal climate change commitment, under which the state of California formalised prior commitments to address its contribution of roughly 1.4% of the world's and 6.2% of total US greenhouse gas emissions. Elsewhere, India is investing in a target 20,000 megawatts of installed solar energy capacity by 2020. Wind power plants currently under construction or in development in the US state of Texas total some 50,000 megawatts, exceeding both the output of 50 coal-fired power stations and the demands of the state's 24 million inhabitants, and therefore potentially enabling Texas (historically famed for its oil wealth) to become a net exporter of renewable power.

The US today is taking what it reports to be a 'common-sense approach to developing standards for greenhouse gas emissions', including regulatory initiatives for both mobile and stationary sources under the Clean Air Act. This includes addressing combined emissions from motor vehicles, promoting a new generation of clean vehicles of all sizes, from small cars to trucks, a Renewable Fuel Standard Program that stipulates a minimum volume of renewable fuel, tightened requirements for the regulation of 'new pollutants', and steps taken in September 2013 under President Obama's Climate Action Plan to reduce carbon pollution from power plants, refineries and cement production facilities. In addition, emissions from large emission sources are being audited more closely under the Greenhouse Gas Reporting Program, while the Oil and Natural Gas Air Pollution Standards aim to cut VOC emissions from hydraulically fractured gas wells. The US is also investing in the geological sequestration of carbon by injecting carbon dioxide (emitted from coal-fired electricity generating power plants, for example) into deep wells to sequester the CO_2 underground indefinitely.

The UK and the EU, like other countries and supranational regions, are also proceeding with legislation to tackle climate change, recognising that this necessarily requires coordinated action by nations around the world. The UK and the EU are party to international legislation that aims to achieve this. The UK's 2006 *Stern Review on the Economics of Climate Change*,[40] though not the first economic report on climate change, became significant as the largest and most widely known and discussed report of its kind, providing considerable leverage on the global agenda. In the review, Nicholas Stern states that climate change is the greatest and widest-ranging market failure ever seen, presenting a unique challenge for economics. The review's main conclusion is that the benefits of strong, early action on climate change far outweigh the economic consequences of continued inaction.

Backed up by innovations in low-energy transport, housing, materials and manufacturing across the world, we are witnessing nothing short of a global revolution in attitudes to carbon and climate, despite a lack of binding set-piece commitments to emissions targets. Notwithstanding consensus about the challenge, we still lack the political courage to set the requisite obligatory targets that would have the added benefit of providing clear signals to businesses, empowering them to innovate new solutions. As a result, we still find ourselves standing at an era-defining crossroads between, in one direction, climate stability and sustainability informing the mainstream of global concern and, in the other, these issues being treated as little more than an afterthought to continuing short-term international economic competition.

The voice of civil society

Activity to address air quality did not arrive purely by top-down imposition. It was substantially shaped by bottom-up public concern expressed, channelled and informed by an increasingly influential and vocal non-governmental organisation (NGO) sector.

Given its fundamental importance, and the visible and immediate health and aesthetic consequences when it is grossly fouled, it should perhaps be no surprise that the air was the focus of the oldest environmental society in the world. The Coal Smoke Abatement Society (CSAS) was founded in 1898 by the London-based artist Sir William Blake Richmond, who wrote a letter to *The Times* in 1898 calling for action and stating that 'the darkness was comparable to a total eclipse of the sun'. The article expressed Richmond's frustration at low light levels in the winter caused by coal smoke. Over the following decades, the organisation was instrumental in introducing in the UK the Public Health (Smoke Abatement Act) 1926 and the Clean Air Act 1956; the latter was instigated as a private members' bill promoted by Sir Gerald Nabarro in the aftermath of the Great London Smog of 1952, and subsequently updated by the Clean Air Acts of 1968 and 1993. As a major consequence of these acts, considerable areas of the UK were declared 'smoke control areas', where the use of certain fuels was either prohibited or allowed only in special appliances. The CSAS was transformed into the National Society for Clean Air (NSCA), addressing wider atmospheric threats as the menace of coal smoke receded.

In the 1970s, the NSCA began to campaign vigorously and influentially on air pollution from other sources, including industry, stubble burning, incinerators (which had Crown immunity), noise and transport (in particular lead in petrol), perhaps reflecting the growing concerns of its predominantly local authority membership. A particularly significant achievement of the NSCA after the Clean Air Acts was the development of the concept of local air quality management, the incorporation of this into the Environment Act 1995, and the production of supplementary guidance documents to assist local authorities in implementing it.

The NSCA was subsequently rebranded as Environment Protection UK (EPUK) in 2007, recognising that its work had spread far wider than just air quality and now strategically included climate change and land quality. By this time, the organisation had also developed a broader membership base, building on a core and still significant membership among local authorities but with substantial contributions from consultants, academics, private individuals and industry.

Another of the pathways through which civil concerns make the tortuous journey to institutionalised societal response is common law, or civil law as it is known in some other jurisdictions.[41] It is hardly surprising to find that various aspects of air pollution are expressed in both statute and common law. Common law in particular has a rich case history relating to odour, flies, dust and noise, as well as to the provision of sufficient airspace to allow chimneys to create a draught.

Shifting cultural views can also become institutionalised by influencing market behaviour, for example with businesses voluntarily de-selecting perceived problematic substances or creating differentiated markets for sustainably sourced forest products, marine fishery and other items.[42] These include, for example, 'climate-friendly' aerosol cans that hit the market following agreements to phase out CFC propellants in the late 1980s. Shifts in common law, expressions of social norms and voluntary market practices can in turn both drive and inform legislative responses.

Emerging issues

New pollutants are constantly appearing as toxic, noxious and other harmful effects are discovered. While many of the issues addressed above relate to impacts on the troposphere (the inner part of the atmosphere that we inhabit), concerns about ozone thinning are stratospheric. However, some emerging issues relate to less tangible chemically based effects.

Noise, too – one of the oldest acoustic pollutants and the subject of a great deal of common law cases as well as statutory constraints – is resulting in novel societal responses. Interestingly, as its chronic effects on wildlife and people become better understood, attitudes are shifting towards the protection of silence rather than the limitation of noise. For example, to reduce noise exposure and preserve the high level of biodiversity offered by naturally quiet areas, the EU Environmental Noise Directive requires member states to identify areas that currently experience very low levels of noise and to preserve these 'quiet areas' from human influence. However, very few member states have done this effectively, because measuring how sound varies in time and space is technically very difficult.

Odour is another form of tropospheric pollution and is variously chemical, biological, or related to other causes. It also has a long history of societal response, but is now gaining increasing attention. In part, this relates to recognition that 'merely' nuisance-based issues can lead to secondary medical concerns such as stress and to economic implications including depressed real estate values, as well as having potential biological impacts. This is forcing a reconsideration of a range of potentially troublesome odour-generating activities such as wastewater treatment, landfill sites, composting activities, the digestion of food waste, sludge handling and sewer gas releases, with associated technological and regulatory innovations.

Among other considerations discussed in Chapter 4: *Abuses of the air* were the insidious and potentially serious implications of light pollution for wildlife and people. A variety of institutionalised responses have been put in place, including a growing body of common law around light trespass and the beginnings of controls on excessive lighting, particularly during antisocial hours, under some jurisdictions. To assist a number of US cities that have developed standards for outdoor lighting to protect the rights of citizens against light trespass, the International Dark-Sky Association[43] has established lighting standards to help redress this common problem and particularly to reduce light going up into the sky, thereby reducing the visibility of stars.

Recognising the value of dark skies to landscape character and tourism, the Northumberland National Park and the Kielder Water and Forest Park in the north-east of England were granted 'dark-sky status' by the International Dark-Sky Association on 9 December 2013; they have achieved 'gold' status, the highest accolade the association can bestow.[44] Dark-sky status means that the night sky is protected and lighting controls are in place to prevent light pollution in the largest area of protected sky in Europe (1,500 square kilometres or 579 square miles). The granting of this status followed a two-year campaign by local institutions and citizens to protect the quality of Northumberland's night sky, as the area is recognised as one of the best places to stargaze in Europe. Measures to secure this valued darkness entail not simply turning lights off, but rather encouraging better lighting using the latest technology. Organisers hope that the award may boost 'astro-tourism' in the north-east of England.

The legal status of air

Compared with terrestrial and aquatic resources, air and the atmosphere have been something of a neglected resource,[45] perhaps in part because they have many attributes of a shared common. The term 'common' lacks a single accepted definition; however, applied to land, it is generally accepted as referring to a resource in either state or private ownership, but to which other people have certain traditional rights to use in specified ways.[46] The services provided by air are in many ways the ultimate global common, conveying water, sound, organisms and their propagules, and gases that may be beneficial, polluting, climate changing or damaging to the air and upper atmosphere. Defining air as a common is rendered more complex by the transboundary, fluid and invisible nature of air, which makes it harder to capture within sovereign jurisdictions, and also more complex to manage.

The extent to which a properly enforced legal regime protecting an environmental resource is maintained is a measure of the value society places on that resource, as legislation effectively represents common social agreements reflecting a complex amalgam of knowledge, public concern and economic interests synthesised into binding protocols through political processes. This review of the ways in which air and its properties have been addressed in legislation, both as a common and as a privately beneficial resource, has a British focus, although most principles are transferable to other jurisdictions. For this legal review, I am particularly indebted to Ben Pontin, Tom Appleby and other colleagues at the University of the West of England in our consideration of 'Air as a common good'.[47]

Legal responses to concerns about air pollution date back 2,000 years to *aerem corrumpere non licet*, which has been the foundation for a raft of statute and common law. Each piece of legislation has marked a small step towards addressing the wider facets of the air and of its close interdependence with human activities. However, these fragmented responses are as yet very far from an integrated approach to air and the many benefits that it confers upon society.

The very fact that air is common to humanity, in terms of its free exploitation, also exposes it to classic instances of the 'tragedy of the commons',[48] such as various forms of pollution. From a legal perspective, the field of common goods is a relatively arcane study; only small areas of the UK in the English and Welsh[49] uplands remain in common ownership and, while commoners have rights against both the freeholder and third parties, these rights do not form part of mainstream legal practice. Roman law (and international law) recognises the concept of something being 'common to humanity'; however, the 'common to humanity' aspect of the ecosystem services provided by air finds an analogue in the 'commoners' rights' to land that were recognised and enjoyed as a form of property right, and that were a powerful force in the long-term protection of common land regardless of the ownership status of the land itself. This, then, begins to address at least some of the ecosystem services provided by the atmosphere, albeit in a diffuse way. Furthermore, while air, like water, is ownerless, some ownership features relate to the services that it provides and that are enjoyed by diverse people; these features include, for example, both common law and the statutory protection of a person's right to enjoy clean air.

Commentators have frequently taken a rather negative view of such common rights. For instance, Vogler[50] recognised conflicting views regarding the 'common' nature of air, stating that:

> One may argue that the atmosphere can also be regarded as a commons, exploited by all yet owned by none. Most significantly the atmosphere has been abused as a 'common sink'. Until relatively recently it provided a completely free waste disposal system for a whole range of

anthropogenic pollutants. It also constitutes the ultimate 'public good', that is to say if resources are expended on improving air quality, it is impossible to exclude people from enjoying the benefits.

In this, Vogler may have been underestimating the ability of the law to provide appropriate remedies to protect common property.

The services provided by air can be brought under legal consideration – and they have been over many years. Examples include the prevention or abatement of air pollution from neighbouring property owners and neighbouring people under the laws of nuisance and public nuisance respectively. As addressed previously in this chapter, a large body of statutory legislation around the world has a long legacy of protecting public interests when proprietary rights have failed. The traditional view of ownership of airspace follows the ancient Roman and common law principle of *cuius est solum, eius est usque ad coelum et ad inferos*: whoever owns the land also owns to the centre of the Earth and to the heavens. The advent of powered flight caused grave problems, with landowners claiming that flight was an aerial trespass (*Bernstein of Leigh v Skyviews & General Ltd* [1978] 1 QB 479). As a result, lawyers have now tended to view airspace in two ways: the lower stratum belongs to the landowner, while the upper stratum is *res omnium communis*, or in common ownership.[51] The lower stratum consists of the portion immediately above the land; interference with this airspace would affect the landowner's reasonable enjoyment of the land and the structures upon it. The Civil Aviation Act 1982 states that this is unlikely to extend beyond an altitude of much more than 500 or 1,000 feet (152 to 305 metres) above roof level, this being roughly the minimum permissible distance for normal overflying by any aircraft. In the higher stratum of the airspace, the landowner has no greater rights than the rest of humanity.

So, while each individual landowner has ownership of the lower stratum of airspace over their property, the air itself is ownerless. However, it is possible for an action to be framed by an aggrieved landowner for pollution of that airspace as a nuisance or a trespass. This leads to some further interesting considerations. For example, the UK seabed, and thus the sky above it, is almost entirely owned by the Crown out to 12 nautical miles, so therefore it probably represents the largest body of publicly owned airspace in the UK, with a single owner who has the power to protect that airspace via tort law.

The beneficial services provided by air may also fall under *salubritas aeris* (roughly translated as 'clean air'), a 'private' common law entitlement of a freeholder or leaseholder.[52] In the case of *William*

Aldred (1610) 77 ER 816, it was reasoned that clean air was a *necessity* of a landholding. As such, pollution of air passing over a proprietor's land constituted a tort that could be remedied by damages and an injunction. However, there has been little or no litigation in which air pollution per se has been the basis of an action under the *Aldred* principle. In the most prominent nineteenth-century air pollution tort of *St Helen's Smelting Co. v Tipping* [1865] 11 HLC 642, the complaint related not to acid gas emissions from the defendant's copper smelting works, but to the injury these emissions caused to the plaintiff's forestry interests (ruining oaks that the plaintiff used for fuel and for the beautification of his estate of 1,300 acres). In the leading twentieth-century action of *Halsey v Esso Petroleum* [1961] 2 All ER 145, the issue with fine particles from the defendant's refinery was that they damaged the paintwork on the plaintiff's car and the clothing on the plaintiff's washing line.

Further interesting legal aspects revolve around the use of the ownerless nature of air itself to produce provisioning services or to 'consume' regulatory services. An example of the former is the extraction of gases, which then become owned and subject to both property law and market forces, whereas emissions into the atmosphere reverse this process from owned to ownerless, at the same time 'consuming' some of the atmosphere's regulatory services to dissipate or break down the gas. Once again, as for water systems, legal custom varies geographically and appeals for mitigation have generally been via the law of tort or general international legal principles such as the UN Covenant on the Rights of the Child.[53] However, if impacts on air impinge upon the enjoyment of its 'common to humanity' properties, the UK courts may be able to use international law as a basis for injunctions or restorative action.

The seminal work 'Property in thin air'[54] quotes various authorities as stating that the upper stratum is *res omnium communis* and, as such, 'open to exploitation by all' and 'belonging to all the world'. These are interesting assertions because they are to some extent contradictory. In one sense, if airspace belongs to all the world, it follows that all the world can then use it as it chooses. But if the airspace is being abused by air pollution or carbon emissions, does it not also follow that all the world can take some sort of legal action to defend that airspace from overexploitation? This could be done in a number of ways, but one particularly useful method is that developed by Sax,[55] who constructed a modern, generalised public trust doctrine around ancient common law theory, relating specifically to the seashore and to navigation. However, the central idea is also relevant to public or common rights to air, particularly in respect of the airspace of the upper stratum, which is acknowledged to be common property. In essence, the public trust doctrine holds that such a common resource is held on trust only and the trustees of the resource should not permit unreasonable damage. In this case, an action could be framed against a common law state for failing in its duties as a trustee to maintain reasonable air quality, causing undue loss to the ecosystem services provided by the air.

In addition to the rights and responsibilities of owners of airspace, the public trust doctrine also has a more esoteric connection with damage to environmental media in general. Although this is a doctrine that has found its most common application in the US, it fits particularly well within the British legal system in which it has its origins. This is because land is not owned within this system; rather, all land is held by the Crown.[56] Proprietors therefore do not have dominion over land as they do elsewhere in the world, and there is always the threat – today in theory more than in practice – that the Sovereign will intervene in property relations to protect the interests of the common wealth. Interventions of this kind ceased during the mediaeval period, when parliament asserted its right to control the exercise of the Crown prerogative. However, there is a good argument in principle that Crown subjects continue to have a right to enjoy the land and its fruits and that it is unlawful for anyone – an individual, a corporation, or a government ministry – to deny that right by means of serious impairment of ecosystem services. It is certainly arguable that air and the atmosphere fall within the rationale of the public trust doctrine, viz.: 'certain interests are so intrinsically important to every citizen that their free availability tends to mark the society as one of citizens rather than serfs'. The notion of society being enslaved by the legally permitted destruction of vital ecosystems is central to this theory.

Further complicating factors include the interaction of environmental media with property, and how this affects enjoyment of the ecosystem services that those environmental media provide. For example, owing to the nature of the media, the services provided by land and water systems (while not entirely distinct from those provided by the air) are often geographically more constrained than air-related services. For this reason, industries, power stations and other infrastructure, including domestic houses, are usually built where they can access land- or water-derived services, whereas air-derived services, which are assumed to be universally accessible and hence overlooked, play less of a part in the siting of a development. This trend is somewhat reversed in the case of the recent growth of wind farms, for which geographic context with respect to high and predictable wind speeds is a major consideration. Nevertheless, where air services have been overlooked, significant air quality problems can arise. One example is Mexico City, where the airshed is partially contained by mountains. In this way, issues relating to urban smog right across the world exemplify how ecosystem services provided by air have been disregarded – with negative implications for the health and beauty of urban development – much in the same manner in which urban sprawl tends to expunge many of the services provided by the watercourses that were, perversely, often instrumental in the founding of settlements.[57] An additional property of air is its capacity for rapid and extensive transboundary movement of pollution, leading to the establishment of protocols such as the 1979 UNECE Convention on Long-range Transboundary Air Pollution.[58]

An interesting conclusion emerging from this discussion is that, while the physical medium itself may be *res nullius* (nobody's property), the airspace and the enjoyment of the services air provides can confer some *de jure* rights both to individuals and to society as a whole for the protection of that air. This may have some bearing on the control of pollution by legislation relating to climate-active, health-related or other gaseous emissions. The main benefit of regarding the atmosphere as a common, then, appears to be to help manage current overexploitation,[59] reinforcing calls for a form of global 'law of the atmosphere' that extends some of the issue-specific principles within the Montreal Protocol and the UNFCCC.[60]

The loss of marketed goods from any environmental medium, such as land contamination due to leaching from industrial sites or the degradation of fisheries, is essentially a matter of private litigation for damages in the UK, US and other jurisdictions where similar models of common law operate. However, in the US there is an additional statutory instrument – the Comprehensive Environmental Response, Compensation, and Liability Act of 1980 (CERCLA)[61] – which enables public trustees to sue polluters. CERCLA was notably applied to address the impacts of the grounding of the Exxon Valdez supertanker on 24 March 1989, resulting in the spillage of 11 million gallons of crude oil into the waters of Prince William Sound, Alaska.[62] Estimates of the cost of the damage to natural resources ran into billions of dollars, based on the application of contingent valuation methods.[63] The US Oil Pollution Act 1990,[64] introduced largely in response to the Exxon Valdez case, required the US President, acting through the Under Secretary of Commerce for Oceans and Atmosphere, to issue regulations that established procedures for assessing the damage to, or the destruction of, natural resources resulting from a discharge of oil covered by the act. The damage categories covered were extensive and included the so-called 'passive' or 'non-use' values[65] of people who do not necessarily intend to make use of the natural resource in question but might value the option of being able to make use of it in the future, or might simply value its continued existence. These principles of 'use' and 'non-use' benefits, whether currently exploited or not and including their associated values, can be interpreted today as relating to services provided by natural resources, and are theoretically also applicable to air. By contrast, environmental harm in England and Wales is regulated by both statute law and common law. Under common law, compensation for damages is assessed in terms of loss of use, which is almost always related to property. While previous studies have tried to find means to address harm to public interests outside the context of 'property',[66] the UK currently lacks any legal equivalent to CERCLA. Consequently, the concept of loss has a clear meaning only when damages are valued, which in practice limits it to markets that exist for what has been or may become lost. The difficulty arises in that many ecosystem services are external to the market, their valuation already noted as a political and policy priority to embed an Ecosystem Approach across societal activities.[67] At present, it is only in the case of exemplary damages, where a deterrent is applied over and above 'use value' losses, that the courts commonly exceed the limitation of awarding damages on traded values, and even this exemplary quantum is not designed to reflect the magnitude of non-use values lost. Thus, liability for damage to the environment is far more limited in UK courts than in American ones, which clearly accept the economic case for including non-use values in assessing the extent of environmental damage. However, with respect to future damage, an injunction that secures the closure of a smelting plant (as in *Tipping*) or a refinery (as in *Halsey*), and its replacement by a land use that is less polluting of the local environment, will have ecosystem service benefits that are wide ranging.

A remedy to this deficiency may result from recent awareness of the value of ecosystem services, whether marketed or not. The UK government published guidance in 2007 on evolving methods for valuing ecosystem services.[68] In 2011, an English government Natural Environment White Paper, *The Natural Choice*, sought explicitly to 'mainstream' these diverse values across societal decision-making.[69] The White Paper promoted 'payments for ecosystem services' (PES) markets (addressed in Chapter 7: *Rediscovering our place in the breathing space*) to achieve environmental objectives, focusing on the potential for markets in biodiversity and ecosystem services and on the environmental valuation problem.[70] Major research programmes such as the Natural Environment Research Council's Valuing Nature Network in the UK[71] and the US-based Natural Capital Project[72] are seeking robust means to capture the diverse economic and non-economic values provided by ecosystems. There remain some concerns about 'putting a price tag on nature',[73] and about the risk inherent in PES approaches of undermining long-established rights to enjoyment of the environment.[74] The Ecosystem Approach explicitly recognises that all ecosystem services provide benefits to different stakeholder groups, and that, whether the service is traded or not, all services have implicit value. Identification and, ideally, eventual internalisation of these 'missing markets' may thus form part of the evolution of environmental valuation and its progressive inclusion into mainstream markets and societal decision-making, including legal remedies.[75]

Missed connections

The breadth of societal responses to threats resulting from the abuse of the air and the atmosphere is impressive. However, hindsight reveals that historical statutory responses have been fragmentary, instituted on a largely piecemeal, issue-by-issue basis in response to specific adverse consequences or incidents – and then only when they became sufficiently acute to trigger public, NGO and ultimately political concern.

Early responses addressed local nuisance and locally acute health risks, including those posed by smoke, smog and a range of more specific air pollutants, as well as noise, dust and flying pests such as flies and scavenging birds, among other issues. Each incremental addition to the body of law marked a small step towards addressing the wider facets of the air, and of its close interdependence with human activities. However, this has produced a tableau of disjointed societal responses rather than an integrated approach to air and the many benefits that it confers upon society. For example, despite the early recognition of win–win opportunities offered by the co-management of carbon dioxide and local air quality pollutants,[76] the norm, at least in the UK, remains that these two issues are managed with little interaction, often at different local government levels.[77] Currently, the EU's LAQM process takes a narrow effects-based approach that focuses on specific gases and fine particulates of concern for human health, and also on exceedances of objective concentrations on a local scale. Although some progress is being made with integration, relatively few win–win outcomes between the two policy areas are being sought proactively. On the other hand, there are many examples of win–lose trade-offs, such as the diversion of traffic from local air quality pollution 'hot-spots', which results in greater overall loads emitted of health-related and climate-active pollutants.[78] This retrospective treatment of the air system is similar to the way in which nature conservation, the water environment and land and landscapes have been treated to date.[79] Fundamentally, the industrialised world perspective that still drives many of the ways in which markets and associated resource use habits operate regards the natural world as a boundless source of resources and a limitless place in which to dispose of solid, liquid and gaseous wastes. Clearly, this varies from a systemic approach that regards and values nature and its processes as vital supportive systems with finite carrying capacity, which therefore need to be factored centrally into governance arrangements. Our collective failure thus far to make this transition is coming back to haunt us once again. Instead, individual issue-based legal protection measures persist, despite a need for an integrated approach across all environmental media and their many societal interactions – a strategy that would be consistent with commitments under the Ecosystem Approach.

Notwithstanding the ineffective co-management with health-related air quality pollutants at local authority scale, international mobilisation around the pressing threat of climate change highlights the progress being made towards a more systemic approach to management. The complexity of pollutant types, the diversity of sources – ranging from direct emissions through to land use, including the drying of wetlands and clearance of forests – and the multinational scales of both problems and solutions require a far more systematic way of understanding the problem. They also demand a more systemic approach to effective responses, although concerted international intent is still often frustrated when it comes to national-level and corporate implementation.

'Cleaning up' on an issue-by-issue basis – and only after adverse impacts have been acute enough and have affected a large enough range of people over a sufficient time period to trigger slow political reactions – has produced fragmented regulatory responses. These responses have often surfaced decades after commercialisation, widespread societal pervasion and the cementing of vested interests around problematic materials and processes. This is like sweeping up an unexpected mess after a series of perceived unfortunate 'accidents', yet remaining blind to the intimate systemic connections between all layers of the atmosphere and other environmental media and all living and non-living things that share those media. This is compounded by the general way in which discount rates are applied in economic evaluations, which tend substantially to underestimate long-term impacts, skewing cost–benefit analyses in favour of the advantages of short-term exploitation, and thereby deferring regulatory responses.[80] This history of issue-by-issue responses to emerging concerns is at best suboptimal, at worst disastrously ineffective.

While policies to control emissions of air pollutants in the UK have focused largely on *direct* impacts on human health, the emerging understanding of ecosystem services provides a broader framework for appraising the potential benefits of controlling air pollution, recognising the additional *indirect* impacts on a variety of aspects of human wellbeing, including health and economic activities. To date, attempts to address the ecosystem service impacts of ammonia abatement have been frustrated by insufficient knowledge of the relationship between dose and response, preventing any prediction of changes in service flows.[81] However, indicative values for the marginal impact of ammonia abatement measures on climate regulation, comparable in magnitude to those for human health impacts, highlight that bringing broader ecosystem service considerations into formerly narrowly framed controlmeasures can demonstrate far greater cumulative societal benefits. Other historically overlooked service implications and pollutant pressures collectively increase awareness of the net costs of abuses to the atmosphere. Notwithstanding methodological difficulties and the uncertainties that they produce, it is clear that adopting an integrated Ecosystem Approach when considering the 'common good' benefits of air, and the implications of harming them, offers the potential for greater recognition of those benefits, including health and wider wellbeing, economic and collateral interests. This reinforces the need – and also the means – to consider the air and the atmosphere as a connected whole, including their connectedness with other media and human interests, and to progressively bring them into the mainstream of policy formulation and broader economic deliberation.

6 | Thinking in a connected way

We live immersed in the medium of air, inseparable from it for our biophysical, metaphysical and economic needs. We share this common breathing space with all things, living and non-living, from the regenerative forest to the smoke-belching factory. All influence our common airspace.

The natural regulatory processes of our home planet are impressive in their magnitude and complexity. However, their supportive capacities are also finite. Stepping beyond natural limits inevitably creates instabilities and risks for all living things, including humanity, its activities and future prospects. As we cannot live outside the laws of physics, chemistry and ecology, the planet cannot sustain our profligacy indefinitely without degradation and ultimate collapse. In this moment in human history, we stand at a crossroads in how we address the conflict between increasing societal resource demands and declining natural 'headroom'.

The straight path ahead – 'business as usual' – will continue progressively to constrain our collective potential to live fulfilled lives. However, other pathways are available, leading towards greater synergy with the benevolent, protective and inherently renewable natural resources and processes of our home planet. One of the key indicators of the path we tread will be how, through the collective actions of all societal interests, we come to regard and treat our common 'breathing space'.

Thinking as if the world were real

Air touches and interacts with everything in the biosphere. It provides raw resources, from oxygen for respiration to carbon dioxide for photosynthesis. It disperses wastes and conveys the aerial phases of water, carbon, nutrient and other supportive cycles. It transports pollen, spores and seeds, scents and pheromones, supporting the flight of insects, birds and other biota. The higher atmosphere shields life from harmful wavelengths of solar radiation, and the actions of life sustain the dynamic instability on which much of this shielding depends, the whole providing a diverse but equitable climate. Air flows through life, and life through air.

Humanity, too, including our economic, cultural and other interests, is a wholly interdependent element of the Earth's biosphere. Air touches and interacts with everything in the anthroposphere. It provides raw resources, from oxygen for breathing and combustion to carbon dioxide for crop productivity. It disperses and breaks down our gaseous wastes, and is a critical medium for the water, carbon, nutrient and other supportive cycles that sustain our health, wealth creation, aesthetic and wider societal interests. It bears our aeroplanes and energises our sails and wind turbines, dampens turbulence, and brings music and speech to our ears. We are shielded from harmful radiation from space, yet equally able to exploit other wavelengths to bounce electronic messages around the globe. We enjoy the planet's various climatic regimes for habitation, agriculture, energy conversion and vacations. Air flows through all human interests, just as humanity forever swims in the sea of the atmosphere, drawing from it a wealth of often unaccounted benefits.

The challenge for global society, given the planetary scale of benefits, threats and necessary responses, is to move more deliberately and rapidly towards systemic thinking and practice.

The journey to a systemic approach

The journey from where we stand today towards systemic thinking and practice, taking account of all ecosystem services provided by the atmospheric system, is necessarily long and daunting. However, we can take heart from the fact that we have already embarked on that journey. A review of how ecosystem services have been internalised into British culture over recent decades highlights how, though not explicitly enacted in those terms, significant progress has already been made.[1] This transition in the UK and much of the developed world takes as its starting point the beginning of the twentieth century, when, as the common saying put it, 'An Englishman's home is his castle'. This reflected the fact that property rights implied the largely unconstrained rights of the (generally male) owner to use that land as he desired. Yet, by the close of the twentieth century, landowners' freedom of action had been substantially limited by a corpus of environmental, industrial, planning and other legislation; a growing body of common case law relating to the impacts of resource use on the rights of other people; incentives to manage the land in certain culturally preferred ways and taxes to dissuade undesirable activities; novel markets for sustainably sourced goods as well as the production of biofuels and feedstock crops partly displacing dependence on fossil resources; catchment management strategies favouring water-sensitive land usage; measures to secure public access; and a range of other changes. Progress over the century may have occurred beyond the span of an average human life, and so has been less obvious to those living through it. However, the telescope of history reveals a broad and profound revolution, which, in historical terms, has been very rapid indeed.

This transformational journey is defined by the recognition that land and other environmental assets are valuable not merely as private property but because they produce a diversity of publicly beneficial ecosystem services, regardless of their ownership status. Beneficial services such as open spaces and fresh air, buffering of storm energy, cleansing of the air and protection of water resources, soil and biodiversity have been recognised progressively and then subsequently valued and institutionalised, be that in statute or common law, in shifting social norms, market differentiation and reform, or through any of a range of other societal 'levers'.[2]

The Ecosystem Approach

Our prior consideration of the multiple ecosystem services – benefits to people provided by ecosystems – that derive from the air and the atmosphere reveals not only their diversity and interdependence, but also how fundamental yet often overlooked they are. The concept and application of ecosystem services have been evolving since the 1980s to support understanding of the natural environment and, in particular, how it supports human wellbeing. As touched upon briefly in Chapter 3: *What does air do for us?*, principles for implementing the evolving ecosystem services paradigm have been developed by the Convention on Biological Diversity (CBD) into the Ecosystem Approach, a term first used in a policy context at the Earth Summit in Rio de Janeiro in 1992.[3] The Ecosystem Approach addresses broader geographical and socio-economic contexts in which ecosystem services are applied, and has gained wide global recognition and commitment, as elaborated in the Annex.

The Ecosystem Approach is a paradigm-changing concept, recognising the tight, if formerly overlooked, interdependence of our needs and activities with the ecosystems that support them. It is also a subversive concept, as so many of today's regulatory, economic, resource use and other industrial and agricultural assumptions, with their associated sets of vested interests, are based in the anachronistic Industrial Revolution paradigm that we can mine from nature endlessly and that the only consequences will be beneficial wealth creation and technological progress. Placing nature and its services at the centre of our thinking across all policy areas is a radical departure, but it is also an essential one if we intend to secure the resources upon which our collective future wellbeing depends utterly.

Yet the invisibility of air, literally and figuratively, is clearly evident in the CBD's articulation of the Ecosystem Approach as 'a strategy for the integrated management of land, water and living resources that promotes conservation and sustainable use in an equitable way'. Without air and the protective atmosphere, this definition is simply not viable, so we have to factor the atmospheric system into our interpretation of the 12 'complementary and interlinked' principles and five points of operational guidance defined by the CBD in order to put the Ecosystem Approach into effect.[4] These 12 principles are reproduced in full in the Annex for those who may not be familiar with them, as are the five points of operational guidance. However, the key principles are restated in Box 6.1 opposite.

The integral consideration of air and the atmosphere is essential to give meaning to, for example: managing living resources as a matter of societal choice (Principle 1); considering impacts on adjacent ecosystems (Principle 3); understanding an economic context that promotes sustainability and internalises costs and benefits (Principle 4); conserving ecosystem structure, functioning and services (Principle 5) by managing them within their natural limits (Principle 6), keeping long-term consequences in mind (Principle 8) and remaining alert to all forms of knowledge (Principle 11); and involving all relevant sectors of society (Principle 12).

Just as the framework of ecosystem services must be regarded as an internally interdependent system of components, so too must the 12 'complementary and interlinked' principles of the Ecosystem Approach. 'Cherry-picking' just a few favoured principles – factoring in ecosystem structure and functioning (Principle 5) and economic context (Principle 4), for example, while overlooking the knowledge and perspectives of all affected constituencies (Principle 11) and temporal scales and lag effects (Principle 8) – is likely to result in inequitable outcomes that are also of questionable long-term value and sustainability, precisely because the systemic context is lost.

Accepting that air and the atmosphere are integral elements in the planetary web of life and in natural cycles, including our access to and exploitation of them, the Ecosystem Approach represents a manifesto for progressive societal transition towards ecosystem-centred

Box 6.1 The 12 'complementary and interlinked' principles of the Ecosystem Approach advanced by the Convention on Biological Diversity[5]

Principle 1: The objectives of management of land, water and living resources are a matter of societal choices.

Principle 2: Management should be decentralized to the lowest appropriate level.

Principle 3: Ecosystem managers should consider the effects (actual or potential) of their activities on adjacent and other ecosystems.

Principle 4: Recognizing potential gains from management, there is usually a need to understand and manage the ecosystem in an economic context.

Principle 5: Conservation of ecosystem structure and functioning, in order to maintain ecosystem services, should be a priority target of the Ecosystem Approach.

Principle 6: Ecosystems must be managed within the limits of their functioning.

Principle 7: The Ecosystem Approach should be undertaken at the appropriate spatial and temporal scales.

Principle 8: Recognizing the varying temporal scales and lag effects that characterize ecosystem processes, objectives for ecosystem management should be set for the long term.

Principle 9: Management must recognize that change is inevitable.

Principle 10: The Ecosystem Approach should seek the appropriate balance between, and integration of, conservation and use of biological diversity.

Principle 11: The Ecosystem Approach should consider all forms of relevant information, including scientific and indigenous and local knowledge, innovations and practices.

Principle 12: The Ecosystem Approach should involve all relevant sectors of society and scientific disciplines.

management, and includes the internalisation of this perspective into all dimensions of societal activity. This is a radical departure from inherited norms – at least for the industrialised world, if not for those in the developing world living what are known as 'resource-dependent lifestyles' (although, of course, all of humanity is 'resource-dependent'). As such, it requires societal structures to adapt to address their cultural dependence on habitats as diverse as wetlands, desert fringes, mountains, catchments, and the shared airspace. The Ecosystem Approach entails valuing the system as an integrated whole in the decisions, policies and innovations in which we invest, and challenging established assumptions, habits, institutional remits and vested interests.

This different world view needs us to pay heed to how the world in which we live works, and to value the future and those who share it over and above the liquidation of resources for short-term gain. This demands a particular type of humility, at once recognising our smallness and our utter dependence on – but as yet far from complete knowledge of – the might and complexity of planetary structures and processes, while at the same time recognising that the scale of our ingenuity and technology empowers us either to unravel this nurturing web or else to redress our historical myopia by working increasingly in synergy with these irreplaceable supportive systems.

From 'outside in' to 'inside out'

The trajectory of human development has been one of ceaseless innovation to improve nutrition, health and longevity, security, shelter, communications and economic prospects. Thus, we have prospered as a species, the majority of us overcoming such biological limitations as insufficient food and water, disease and predation. Our technological advancement – from energy supply and exploitation to food production from land and sea, communications and transport – is nothing short of miraculous. However, it is only in relatively recent decades that society has been forced to confront just how antagonistic many of the technologies pervading modern lifestyles are to the supportive processes of the natural world with which we co-evolved, and which remain vital for supporting our prospects into the future.

In large measure, societal responses to threats to air and the atmosphere, as well as to other supportive environmental media and their wealth of biodiversity, have been piecemeal and, above all, retrospective, occurring only when acute consequences have become apparent. As we have seen, responses to problems arising from abuse of the airspace have included the creation of a body of statute and common law, quasi-legal protocols, market adjustment or differentiation, incentives and taxes, and a range of other modifications of societal norms. Yet development assumptions still bear the legacy of our industrial past, with environmental consequences generally considered only as an afterthought to technical innovation and economic gain, and then often implemented only as a means to mitigate impacts that have manifested acutely enough to demand an institutionalised response. We still think from the 'outside in', in terms of retrospectively addressing the consequences of decisions that were taken without their 'fit' with ecosystems being a central consideration.

The Ecosystem Approach is fundamentally about establishing that 'fit' from the inside out. It is about recognising that all societal activities have dependencies on the natural world, as well as there being a wide range of potential ramifications for supportive ecosystems and all who share and benefit from them. As highlighted by the transition in the UK throughout the twentieth century, we are making halting progress in addressing more of the services of nature. However, the goal is surely to consider all ecosystem services as a fully interdependent system at the point of decision-making, otherwise there is a high likelihood that we will merely perpetuate our legacy of unintended negative consequences as a result of overlooking those interdependencies. For guidance, we now have the systemic ecosystem services framework and the wider context-setting Ecosystem Approach to guide decisions and to ensure that they best serve optimal human benefit, economic efficiency and the safeguarding of productive ecosystems. This is the essence of the transition to thinking, planning and innovating from the 'inside out'.

This means that all facets of the supportive and protective air and atmospheric system should be considered centrally in planning, innovation and policy- and decision-making. This is not merely ethically and environmentally responsible, but also economically efficient, as it optimises the benefits to all in society as well as averting unintended harm to ecosystems and the associated liabilities and costs of that harm. This transformation, from 'outside in' to 'inside out', is not just an adjustment but a wholesale paradigm shift from implicit post-industrial assumptions underpinning much of the developed world and its legacy of technologies, regulations and economic drivers. The threat of declining environmental integrity and supportive capacities for the burgeoning global population means that embarking on this journey is vital if we are to avert an increasingly impoverished and conflicted future.

So how might we do things differently?

A range of management arrangements that address aspects of the air environment highlight the opportunities that would have arisen for greater synergy had an ecosystem services perspective been adopted. There is no implied criticism of the historical trajectory of these schemes; many of them were established before the genesis of contemporary systemic approaches to understanding and management. However, each of the following case studies illustrates transferrable opportunities and shows how greater synergy and net societal benefits could be achieved in future, including the avoidance of unintended negative consequences.

Synergies between the management of local air quality and climate change In the previous chapter, *Managing our impacts on air*, we saw how the EU Air Quality Management/Local Air Quality Management (AQM/LAQM) process has been framed and implemented around a narrow, locally defined, effects-based approach for specific substances already implicated in acute health impacts. There is no guidance about how other substances that may potentially pose a problem, or for which the effects are unknown, should be handled under the LAQM framework other than the more general EU and global presumption in favour of the 'precautionary principle', the practical implementation of which is fraught with difficulty. Other impacts, even from known problematic substances such as those implicated in climate change or ozone depletion, are not addressed within the AQM framework, even though 'air quality' gases on a local scale largely derive from similar sources.[6]

In theory, climate change strategies could be integrated into aspects of the LAQM process to facilitate local authority-driven reductions in carbon emissions. However, in practice, action to address climate change generally occurs in parallel rather than in synergy with measures to address health-related air pollutants. In the UK and much of the rest of the world, these two sets of problems are handled discretely under different strands of legislation, government guidance and non-statutory agreements, and regulations are often enacted by different management tiers in local authorities, with little coordination between them.[7] This occurs despite the fact that the LAQM process involves methods, skills and collaborative networks that could readily be exploited as a key element of an effective carbon management framework at the local level. This disconnect between management regimes represents a significant missed opportunity in legislation and policy for co-management, due to the wide range of both 'win–win' opportunities (such as emissions reductions addressing both health-related and climate-active gaseous impacts simultaneously) and 'win–lose' trade-offs (such as traffic diversions to reduce the local pollution of health-related gases that may increase overall volumes of emissions of these substances as well as climate-active emissions that have an impact on a global as well as a local scale). From a whole-systems perspective, environmental policies for air quality management and climate change should no longer be treated in isolation due to their potentially synergistic or antagonistic relationships.[8]

Beyond these climate-related concerns, all other ecosystem services are currently omitted from measures put in place to control local air quality. Each ecosystem service represents a contribution to net societal value that could be generated by co-management actions, or else a risk in terms of unintended negative consequences that undermine human wellbeing, system resilience and net value. Relatively little research has been carried out on the wider ecosystem service benefits of air quality control measures, although (as mentioned in Chapter 5: *Managing our impacts on air*) a study of the marginal impacts of ammonia abatement measures on climate regulation found indicative values comparable in magnitude to those for human health impacts. This highlights how bringing broader ecosystem service considerations into current, narrowly framed control measures can yield increased net societal benefits.[9]

At present, many actions put in place to address localised exceedances of one or more 'problem' air quality pollutants in a narrowly framed way simply result in the redistribution of pollutants outside the locally identified 'hotspot', and are likely to exert a range of negative impacts on other services. Some of these potential, but often poorly understood, wider-scale impacts include the effects of pollutants such as ozone on agricultural productivity, and the possible contribution of gases containing nitrogen to eutrophication and potentially problematic changes in habitats and other natural resources.[10] Further unaccounted impacts that can arise from diverted traffic and associated emissions include those on aesthetic quality and property values,[11] as well as on tourism potential and the quality of air used for industrial purposes. Strikingly, the current narrow and fragmented policy environment means that these potential antagonisms are simply not considered in decision-making. Given financial pressures to cut costs by ignoring or failing to tackle 'externalities' to the market and legislation, wider negative consequences seem an inevitable result of the current, narrowly framed regulatory and market drivers.

By comparison, an Ecosystem Approach requires consideration of the full suite of ecosystem services, addressing whole socio-ecological systems and the many stakeholders on different spatial and temporal scales likely to benefit or suffer from management changes. This is not only a more inclusive approach, recognising transparently the likelihood of adverse impacts and identifying stakeholders who should be considered and, ideally, involved, but also one that provides a basis for innovation in the synergistic delivery of multiple benefits across policy areas.

Practical examples of measures that both address LAQM priorities and also deliver wider ecosystem service benefits include building walking routes and affordable public transport systems into cityscapes, thereby servicing some local mobility needs and potentially averting vehicle movements and associated emissions. The promotion of mixed developments, in which employment, retail and living spaces are juxtaposed, may also reduce the need for journeys; this highlights the need for integration between policy areas that do not traditionally consider air quality implications. There are clear potential economic benefits from policy integration, providing multiple and diverse 'win–win' outcomes across policy areas and optimising net societal value. With strategic planning, the co-benefits of considering ecosystem functions in urban design can include visual and noise screening within developments as well as places for wildlife, carbon sequestration, local flood management and recreation as well as the aesthetic enhancement of local sites, which have the potential to elevate real estate values.

This potential for synergy was recognised in the UK's *Stern Review on the Economics of Climate Change*[12] which states, for example, that:

> Policies to meet air pollution and climate change goals are not always comparable. But if government wishes to meet both objectives together, there can be considerable cost savings compared to pursuing them separately.

The same principle applies across all policy areas and the ecosystem services associated with them, including those services currently not central to decision-making but that may nevertheless be impacted along with their beneficiaries.

The co-benefits and net cumulative value of a more coordinated climate change policy are becoming clearer; international collaboration to limit climate warming to 2°C could provide additional health, ecological and economic benefits estimated to reduce global expenditures on air pollution control in 2050 by €250 billion across EU-27 countries.[13] Climate change mitigation alone, although essential, appears expensive, yet if the co-benefits of associated reductions in air pollution are also accounted for, there may be substantial net economic savings. This represents a challenge to the traditional narrow focus of policy-makers and markets, and also demands more information, yet is a key stepping stone to a more sustainable approach to investment and decision-making.

Turning unintended climate impacts into co-benefits in wastewater treatment Across much of the world, the water services industry is constituted largely on a technocentric basis. Financial and environmental regulatory drivers and associated ring-fenced budgets largely reflect an electromechanical paradigm of technical treatment of water abstracted for public supply and of reduction in pollutant loads in wastewater before discharge back into the environment.

For example, since privatisation of the water industry in England and Wales in 1989, substantial investment has been mobilised to improve the treatment of sewage, resulting in significant improvements in the quality of water in rivers and the coastal zone.[14] This progress has coincided with a range of increasingly stringent requirements under various strands of legislation to address the quality and various uses of the water environment. Some key legal drivers, mainly emanating from Europe, include the 1978 EU Freshwater Fish Directive (2006/44/EC), the EU Shellfish Waters Directive (2006/113/EC), the 1991 EU Urban Waste Water Treatment Directive (91/271/EEC) and the 2000 EU Water Framework Directive (2000/60/EC). Additional measures also address the impacts of sewerage and wastewater treatment infrastructure on odour and other aspects of local air quality. However, these substantial improvements in water and local air quality were largely achieved through the increased energy inputs needed by advanced wastewater treatment technologies, including activated sludge and trickle filtration, with their associated increases in treatment chemicals and their substantial transport and supply chain 'footprint', as well as increased waste generation, vehicle movements and other civil disruption to service treatment facilities. Furthermore, methane generated from wastewater plants is highly climate-active.

Increasing awareness of climate change and its associated risks, allied with emerging legal requirements to halt the growth of emissions, is bringing excessive energy use by all sectors of industry and society under closer scrutiny. In retrospect, the single-discipline focus on improving water quality, sometimes with local odour benefits, was achieved at net harm to the atmosphere through substantially increased emissions of climate-active and other health-related gases in associated energy generation. In addition, pressure was put on natural resources and on the people involved in the supply chain of chemical inputs, habitats were lost to landfill for waste chemicals, and there was increased disruption due to vehicle movements. The EU Water Framework Directive (WFD) represented a regulatory evolution in that the means by which its principal target – achievement of 'good ecological status' – is to be achieved are not specified. However, it lacks a wider, systemic consideration of impacts across the full spectrum of ecosystem services that are likely to arise from 'programmes of measures' put in place to meet WFD requirements.

The need to address water quality improvements in wastewater treatment in more environmentally benign ways has been a relatively recent innovation. This includes technologies such as constructed reed beds, which break down organic, nitrogenous and other waste materials using natural processes, thereby producing reduced amounts of climate-active and other polluting gases.[15] Systemic perspectives also recognise the importance of attenuating run-off and associated pollutants at source through measures such as sustainable drainage systems (SuDS), before flood and other wastewater can accumulate, together with associated loads of mixed pollutants, in surges that may enter watercourses or sewerage systems.[16]

Fully integrated solutions include integrated constructed wetland (ICW) systems, which explicitly target 'landscape fit' and the production of multi-benefit outcomes across all ecosystem services, significantly including those that decontaminate wastewater.[17] The integrated control of both point sources and diffuse inputs of pollutants across catchment landscapes is thus receiving greater attention, at least where these progressive measures have been implemented, and is resulting in benefits to the atmosphere. So, while the air quality impacts of water treatment options have been overlooked to date – except perhaps for odour and sometimes for flying insects, which are often a source of complaints from nearby landowners and land users – broader atmospheric impacts should be integrated progressively with combined land and water management in a more systemic approach to all environmental media and their many beneficiaries.

The ecosystem services framework helps identify potential innovations, including cost-efficient multi-benefit approaches, over and above averting any unintended harm in technology choices.[18] Obstacles to the widespread uptake of these multi-benefit approaches are mainly institutional, including a legacy of regulations framed around narrowly defined disciplinary outcomes, associated restricted budgets allocated to achieve single-issue goals, fragmentation of responsibilities for delivery across institutions and departments, risk aversion, and fixed assumptions by regulators, consultants and decision-support models.[19]

Achieving multiple benefits in the protection of water quality It is not only at the 'dirty water' (wastewater treatment) end of the water service industry that progress is being made in using ecosystem processes to attain more sustainable and beneficial outcomes. The technocentric, single-issue focus that has been the dominant approach to wastewater management to date across much of the world, and that has been blind to many potential unintended negative consequences for non-focal services, is as dominant at the 'clean water' end of the industry. Consequently, many water service providers are incentivised by the regulatory and financial environments to build, own and operate 'hard engineering' assets to abstract, store and transfer raw water, and to treat and distribute it. Until relatively recently, little emphasis had been put on protecting the raw resource at source, with some major exceptions: in eastern Australia, the scarcity of water means that there has long been a focus on the protection of vulnerable upper catchments serving water supply reservoirs. However, this technocentric paradigm is beginning to change.

In reality, constrictive institutional walls governing water resource exploitation have been dissolving, albeit slowly, over several decades. From the earliest stages of legislative controls on industrial pollution, some reviewed in Chapter 4: *Abuses of the air*, the focus was on the nature of the polluting industry. Progressively since the late 1970s, the focus shifted towards the quality of receiving waters and some of their uses, including their potential for abstraction for public water supply, industry and irrigation. Arguably, the greatest step change in regulation since then is exemplified by the WFD, with its goal of attaining 'good ecological status' in all water bodies. However, voluntary initiatives undertaken under the leadership of some water service businesses and promoted, often in partnership, by Rivers Trust non-governmental organisations (NGOs) are breaking the paradigm by focusing on improving the quality of raw water and tackling contamination at source, rather than through technical treatment post-abstraction.

The Upstream Thinking programme, for example, is a global exemplar of managing landscapes that yield water.[20] It reinvests a proportion of revenue from the customers of South West Water,[21] the regional water utility for the South West of England, into grants for infrastructure, subsidies and advice to targeted farm businesses in rural catchment areas upstream of water abstraction points. Measures such as fencing stock out of vulnerable streams and wetlands, reducing pesticide and fertiliser inputs and improving storage facilities, separating clean roof water from dirty yard water, providing shelters to over-winter stock and improving slurry storage facilities and spreading practices have win–win benefits. Not only do they enhance the hydrology and quality of watercourses, they also contribute to farm incomes by, for example, reducing foot disease and parasite transmission in stock, helping prevent stock straying, and averting the loss of lambs dropped on waterlogged land. South West Water acts in partnership with the Westcountry Rivers Trust (WRT),[22] an NGO established in the mid-1980s largely in response to the frustrations of landowners on the Tamar and other West Country river systems about the lack of action to halt the decline of salmon stocks or the ravages of diffuse pollution. Since its inception, WRT has developed close relationships at farm-business scale, and so South West Water was quick to see the value in partnering with WRT for delivery of the Upstream Thinking programme.

Ofwat, the Water Services Regulation Authority,[23] which is the statutory financial regulator of the water industry in England and Wales, accepts an estimate that the 2010–15 Upstream Thinking programme is likely to achieve a massive 65:1 benefit-to-cost ratio relative to traditional downstream electromechanical treatment of contaminated water. However, this provides only a limited view of the net co-benefits of measures promoted by Upstream Thinking. Reflecting some of the 'siloed' nature of regulation and budget allocation referred to above, Ofwat recognises only water service industry benefits as admissible in cost–benefit analysis. However, the advantages of a focus on protecting or restoring the natural functions of catchments extend way beyond this, reflecting a far wider socially beneficial 'basket' of services, including, for example, carbon sequestration on better-managed fields and in buffering wetlands, and more complex vegetation in protected zones – this not only benefits biodiversity and fish recruitment but also better regulates air pollutants, microclimates and wind-driven storm damage. Reduced dependence on traditional 'hard engineering' treatment intensity also considerably saves indirect emissions to the air from energy inputs as well as in the supply chains of treatment chemicals, haulage on- and off-site and gaseous emissions from landfilled waste materials. Benefits to air and the atmosphere are part of a package of co-benefits to other environmental media, social needs and costs through working with natural processes.

In England, this initially NGO- and industry-driven focus on catchment ecosystem services as critical core resources has significantly influenced the policy environment. For example, the government's Catchment Sensitive Farming[24] programme of farm adviser visits closely emulates the Rivers Trust's catchment improvement projects. The UK government's Making Space for Water[25] programme of natural flood-risk management now looks to use, restore or emulate natural catchment hydrology to address flood risk at source rather than relying on expensive flood defence structures around downstream assets at flood risk. Furthermore, the 2013 Catchment-based Approach[26] seeks greater stakeholder participation and priority-setting in the ways in which catchments are managed. These represent steps towards a systemic approach, albeit one still constrained by various entrenched vested interests, assumptions and regulatory and budgetary myopia.

These English examples are far from isolated, both across the UK and in the wider world. Other examples in Europe include project ALFA[27] (Adaptive Land use for Flood Alleviation), which brings together partners from a number of European countries to explore and implement new ways of protecting people from flooding by adjusting land use to create capacity for water storage or to abate flooding peaks. Multinational industries such as Nestlé Waters are recognising that partnerships with farming interests to promote catchment management are not only the most cost-efficient but perhaps the only workable methods available to protect the springs producing natural mineral water for its Vittel[28] and Perrier brands. Uncosted co-benefits of protecting the springs include lower emissions to the air from the intensive land use practices also responsible for water contamination. On a grander scale, the development of the New York City water supply, covered in great detail in my book *The Hydropolitics of Dams: Engineering or ecosystems?*,[29] also addresses water supply issues through an urban–rural partnership. This promotes and funds catchment protection, resulting in water supply benefits that include vastly less energy and associated chemical inputs for downstream treatment, while also contributing to the uncosted retention or rebuilding of carbon in soils. This ecosystem-based approach is taking root in New Zealand too, in the context of forest protection as the nation undergoes a transition to a more urban-based economy; policies to subsidise forest and landscape protection serve a number of linked benefits, including carbon sequestration but also water quality protection, nature conservation and maintenance of the traditional livelihoods and culture of indigenous Māori landowners.[30] (This is touched upon in a little more detail in Chapter 7: *Rediscovering our place in the breathing space.*) The Māori ethic of stewardship, *kaitiakitanga*, demands a balanced approach to safeguarding the legacy of previous generations, the needs of current generations, and opportunities for future generations – a fair articulation of what many of us understand by the term 'sustainability' but also one that highlights how the protection of ecosystems and their processes is beneficial for safeguarding not only air and the atmosphere, but also water, land, biodiversity and people. In China, the Sloping Land Conversion Policy, initiated following devastating floods along the Yangtze River in 1998, is signalling a significant change in the country's approach to ecosystem management, contributing not merely to the protection of water resources and the retention of sediment on land but also to a reduction in aeolian soil loss and its sequestered carbon content, which contributes to the re-emerging vitality of the Loess Plateau.[31]

The shifting paradigm of water resource protection and flood management offers many significant lessons, but perhaps the most pertinent for this consideration of air and the atmosphere is the following. A focus on regenerating the structure and functioning of ecosystems is central to such important services as carbon retention and sequestration, 'scrubbing' of contaminants from the air column, and reducing dependence on the electromechanical engineering solutions that are the source of so many aerial impacts from energy and chemical inputs and waste generation.

The Montreal Protocol and beyond The 1987 Montreal Protocol, reviewed in Chapter 5: *Managing our impacts on air*, was a milestone on a geopolitical scale in the recognition of and management response to threats to the planet's ozone layer resulting from the release of chlorofluorocarbons (CFCs) and a range of other ozone-depleting substances. The treaty required parties to reduce their use of regulated substances by half by 1999 relative to 1986 levels, with subsequent amendments phasing out certain specific substances by a range of target dates. Yet, notwithstanding these significant successes, a principal limitation of the protocol was its substance-by-substance specific focus, rather than taking a systemic approach to all substances with known or suspected ozone-depleting properties. This resulted in perverse consequences for both ozone-depleting substances as well as those with wider environmental effects. For example, one of the immediate results of agreements to phase out CFCs was their substitution with HCFCs (hydrochlorofluorocarbons) on the refrigerants market, as

HCFCs required no changes to refrigeration technologies.[32] Albeit to a lesser extent, HCFCs share some of the ozone-depleting properties of CFCs, but they are also more climate-active.

Had a systemic Ecosystem Approach been adopted, substances would have been assessed strategically in terms of their total impacts on all services, including outcomes for human health, climate change, ozone depletion and other aspects of wellbeing. This would have resulted in a more rapid phase-out of ozone-depleting gases thanks to the identification of necessary technology shifts, rather than marginal alterations perpetuating current problems. Such a strategic transition would also have proved to be more cost-beneficial in the long term, leading to both the development of new refrigerant chemicals with a longer product life as well as less wasted investment in 'stranded assets' when the problems inherent in HCFCs and other alternatives became apparent and these substances in turn were scheduled for phase-out.

Control of fine particulates The preceding case studies take a present-day view of historical developments to highlight how an ecosystems perspective might have led to different and more beneficial policies and actions. By contrast, management responses are still being formulated to the issue of how best to manage airborne fine particulates, including $PM_{2.5}$ and PM_{10}, which represent a respirable fraction implicated in a range of cardio-vascular and cardio-pulmonary diseases.

There are many options for managing $PM_{2.5}$. For example, increasingly fine filters could be fitted to vehicle and other emission sources, although these would inevitably affect efficiency as well as generating hazardous waste. Other options include changes in combustion and other fuel technologies, the rerouting of traffic and other emission sources away from built-up areas, low-emission vehicles and pedestrian- or delivery-only urban zones, and an increase in urban tree cover to help 'scrub' particulates from the air. The most easily implemented of these options to address local exceedances of concentrations set out in air quality standards are similar to those identified under the LAQM framework. These include, for example, the rerouting of highways, but while this may address PM_{10} 'hotspots', it may also create disruption, the redirection of pollution, increased climate-active gas emissions, reduced property values and aesthetics elsewhere, and a range of other 'win–lose' outcomes for ecosystems, and for their services and diverse human beneficiaries. If planning proceeds on the basis of impacts on the whole socio-ecological system, reducing local concentrations in ways that do not cause net harm to other services and their beneficiaries (i.e. the extended Pareto principle[33]), then novel options might be identified that can deliver a wider range of societal benefits.

For example, trees are known to be effective in removing particulates from the air column.[34] Increasing tree growth and associated non-paved areas in the urban environment is also known to produce a range of net benefits including improved hydrology (flood control and groundwater replenishment, for example), sound buffering, carbon sequestration, improved microclimate (including breaking down 'heat islands'), access to green spaces (often enhancing property values as well as providing an opportunity for healthy exercise), and space for biodiversity – all of which yield substantial and quantifiable benefits.[35] When combined with other cross-disciplinary measures, such as the integration of urban tree planting with SuDS or other 'green infrastructure', perhaps allied with walkable routes and biodiversity corridors, major 'win–win' outcomes may be achieved across policy areas from single, cost-effective interventions. A more strategic option is, of course, to promote a transition to cleaner fuels so that fine particulate matter is not emitted in the first place; this also spans multiple policy sectors, and an equally diverse range of deeply vested interests in the technological status quo that may argue for simple traffic diversion or exhaust gas control measures to perpetuate their interests. There is a stark choice open to policy-makers: repeat outmoded, single- or few-discipline approaches likely to result in inadvertent 'win–lose' outcomes, beneficial to a minority but undermining the wellbeing of ecosystems and the many who benefit from them, or take an Ecosystem Approach to innovating and show strong leadership in promoting integrated 'win–win' solutions of optimal value to all stakeholders.

Historically, pollution control has been founded largely on agreed standards reflecting 'threshold' concentrations of known pollutants adjusted by a variety of safety factors.[36] However, a growing body of evidence suggests that fine particulates are 'no threshold' pollutants, since there is no clear level below which they have no harmful effects.[37] As other epidemiological evidence builds, breaking with entrenched assumptions that all pollutants have threshold concentrations, it is likely that further categories of 'no threshold' pollutants will be identified. The discovery of formerly unsuspected impacts over recent history – 'no threshold' pollutants, endocrine disruption, teratogenicity (the causing of birth defects) and the stimulation of cancers are just a few examples – has become an issue of mounting concern since the 1990s. It is one that provides a compelling reason for taking a more integrated Ecosystem Approach to the atmosphere, protecting its status and services on a precautionary basis for long-term benefits rather than responding reactively to tackle specific pollutants only as adverse consequences become evident.

Energy strategy Knowing what we know now, would we have made the same decisions about energy strategy that we made in the past? If we had acted on what we knew at the time, should we have made different choices more recently? And do current decisions on energy policy accord with current knowledge of the consequences, and indeed with stated commitments under theoretically binding protocols and policies?

Climate change had hardly emerged as a scientific concern, let alone a political issue, when acid rain found its way into political consciousness. The predominant response to more stringent sulphur dioxide (SO_2) emission standards was the implementation of flue-gas desulphurisation (FGD) to scrub up to 95% of the SO_2 content of exhaust flue gases generated by fossil fuel power plants and some other industrial processes with significant emissions of oxides of sulphur. FGD comprises a suite of techniques, including wet scrubbing with slurries or dry sprays of alkaline sorbents such as lime and limestone, and the recovery of sulphuric acid from sulphur-rich emissions. FGD implementation first occurred in the 1930s to control the more acute impacts of power plant emissions, but became far more widespread with rising political concern in the 1970s. With hindsight, however, FGD is a single-issue fix to a narrowly framed problem, dealing with one emerging concern without asking questions about strategic goals. What, for example, are the risks associated with supply chains of alkaline substances used to immobilise the SO_2? And what other issues arise from the need to transport and dispose of large volumes of solid waste? More importantly, knowing what we know now, does the lock-in of investment inhibit more strategic thinking about subsequently emerging energy-related issues, such as climate change?

Moving ahead to the 2010s, we are seeing considerable investment in carbon capture and storage to inject carbon-rich waste gases into underground strata. Superficially, this makes sense in terms of reducing climate-active emissions. However, in the light of international commitments to a move to renewable energy sources, this 'end-of-pipe' fix is clearly a far from strategic approach, locking in capital to perpetuate existing problematic technologies. Furthermore, taking account of contemporary knowledge about wider systemic impacts, focusing once again on fixes to narrowly defined issues makes far less sense, particularly as this will result in the lock-in of capital that is no longer available for decarbonised innovations.

Decisions by the US and the UK, among many other developed nations, to give consent to and invest heavily in fracking (hydraulic fracturing) to liberate and exploit cheap shale gas also fly in the face of stated commitments to decrease our dependence on fossil carbon-based energy. The fact that it will contribute to overall greenhouse gas emissions is the only certainty. The massive current expansion of fracking is occurring despite substantial unknowns about its implications for wildlife, geological stability, the use and eventual fate of large quantities of water and associated additives, implications for local property values, and the wider consequences stemming from cheap exported coal addressed in Chapter 5: *Managing our impacts on air*.

Across the world, various subsidies have been put in place to help bring renewable energy conversion technologies into the mainstream and to make them cost-competitive as economies of scale are realised. This is a welcome incentive to help accelerate the kinds of technologies that will be necessary in a more sustainable, or at least a more resource-constrained, future. The UK, for example, has been among the leaders in this field with its feed-in tariff. However, part of the response of UK government to the global economic slowdown of the early 2010s was to cut back substantially the scale of these incentives as a cost saving. This is clearly short-termism, opting for immediate cash savings while ignoring the many costly, long-term externalities associated with the perpetuation of carbon-intensive energy generation. It will also quell the stimulus of technology development, and the realisation of patent rights and other advantages that would have been of real benefit to the British economy in a future that will necessarily require these technologies.

The UK approach to both FGD and incentives for the innovation and implementation of renewable energy technologies contrasts with the situation in Germany, which has made a strategic commitment to achieving a carbon- and nuclear-free future. Germany set out a zero-carbon road map in 2010,[38] releasing a major report by Umweltbundesamt (Germany's Federal Environment Agency) to signal that it planned to decarbonise its electricity network. This report described how the country could phase out fossil fuel power plants and replace them with existing renewable energy technologies such as wind turbines and solar panels to achieve an entirely renewable energy supply by 2050.[39] German greenhouse gas emissions had already decreased by 17.2% between 1990 and 2004, and Germany actively promotes government carbon funds and supports multilateral carbon funds with the aim of purchasing carbon credits (a tradable certificate or permit representing the right to emit carbon dioxide or another greenhouse gas expressed as a carbon dioxide equivalent, tCO_2e), as well as working closely with major utility, energy, oil, gas and chemicals conglomerates to maximise the cheap purchase of greenhouse gas certificates (certification of carbon emissions or savings).

Germany is also a leader in innovations and patents in wind and solar generation, positioning it strategically not only for better environmental performance but also for greater resource efficiency and economic competitiveness. In addition, Germany is already well established as the world's largest generator of solar energy and second-largest producer of wind energy after the US, generating 16% of its energy from renewable sources, with further substantial increases in renewable capacity planned over the next decade. Germany is the largest country so far to set a target of becoming carbon neutral, but a number of other countries, including the Maldives, Norway, New Zealand and Costa Rica, have also made this pledge.

Transport A similar crossroads between old technology and new approaches is observed in the transport sector; some of the issues have already been aired when considering the management of air quality and of $PM_{2.5}$ and PM_{10} matter. Do we stick with the model we have and try to 'green' it with, for example, platinum-rich catalytic converters, although this would involve all sorts of metal pollution, supply chain security and footprint issues? Or do we take an eco-efficiency approach by investing in 'lean burn' engines to reduce the overall volumes of emissions? Or are there other models, such as electric vehicles, that have zero emissions at the point of use, although they have to be charged from generally remote sources of uncertain provenance?

The mantra of 'reduce, reuse, recycle' is commonly mooted when addressing resource use problems, but reduction, or indeed complete elimination, is rarely the starting point. Yet why do we need as much transport? Are there more strategic ways of avoiding transport dependency? As discussed above, mixed-use development could make a major contribution to averting short vehicle journeys, by planning safe and attractive walking, cycling and other low-energy transport links between working, residential and shopping spaces.[40] These links also could be vegetated to enhance their visual appeal as well as their capacity to absorb pollutants, noise, visual blight and surface floodwater, sequestering carbon and providing green spaces for amenity and wildlife. Investment in videoconferencing and teleworking could also help reduce current and rising levels of daily commuting, and the cost savings of this for enterprises are also immediate.

Government subsidy for mass transport is seen as expensive in many developed nations, with high fares from privately operated transport modes dissuading many potential passengers from using possibly more sustainable options in favour of the private car. Yet there are many examples around the world of how a greater focus on wider economic and sustainability benefits can promote investment in public transport. Perhaps the most famous is the frequent and rapid bus and integrated public transport system of Curitiba, Brazil, which is cheap, comfortable and attractive, with conveniently situated stations, and which plays a large part in the congestion-free streets, pollution-free air and wider liveability for the 2.2 million inhabitants of greater Curitiba.[41] Given the often substantial 'food miles' associated with the transport of food from field or sea to plate, further sustainability benefits could flow from locally sourced food were there the courage to look and plan across traditionally isolated policy areas. Sustainable food is being pioneered in the English city of Bristol, where a food policy council provides support and development help.[42]

However, we are clearly still a good distance from this goal of connected thinking, as presumption in favour of individual transport systems predominates in much of the developed and developing world, with ever greater investment in road infrastructure and the fuels that generate a substantial proportion of the very climate change that these nations are ostensibly committed to addressing. This contradiction finds its apotheosis in the ways in which obligations to include a proportion of biologically based material in road transport fuels in Europe are being addressed. Drafted with the best of intentions to reduce emissions of climate-active gases, the EU Renewable Energy Directive[43] includes mandatory sustainability criteria for biofuel use in transport fuel, under which suppliers of liquid or gaseous biofuels derived from crop-based feedstocks, wastes or non-agricultural residues apply for tradable renewable transport fuel certificates. There are stipulations that biofuels must not be made from raw materials obtained from land with high biodiversity value, or from land with high carbon stocks such as forests or undrained peatland. However, in addition to its widespread use in food products and cosmetics, palm oil is increasingly being used for the production of biodiesel, with devastating environmental consequences as tropical rainforests and peatlands in South East Asia are being torn up to provide land for lucrative oil palm plantations. Consumption of palm oil is predicted to more than double relative to 2000 levels by 2030 and to triple by 2050, with the amount used by the biofuels industry expanding rapidly.[44] The irony is that attempts to reduce the impact of climate change by imposing obligations around the world, from China to India, Europe and the US, to replace a proportion of fossil fuels with biologically based alternatives appear to be making things significantly worse: the process of clearing forests and draining and burning peatlands to grow palm oil releases substantially more carbon emissions than burning fossil fuels, as well as spelling disaster for local communities and biodiversity.[45] International efforts are being made to address some of the high-profile concerns around the sustainability of palm oil, significantly under the Roundtable on Sustainable Palm Oil.[46]

The key lesson for transport, as indeed for many other issues, is that single-issue techno-fixes without broader thinking about truly sustainable innovations are likely simply to generate new, unsuspected or disregarded problems. And while these fixes are generally profitable for a minority, they result in dire externalities that are shared by many others, including future generations, thwarting real progress towards sustainability while cementing the status quo with all of its various vested interests.

Lessons emerging from case studies

Several key lessons emerge from this brief consideration of how things could be done differently, knowing what we know today about how air, the atmosphere and the wider biosphere works, their vulnerability to what we are doing to them, and the need to take far better care of them in the future. These lessons can inform us when addressing the pressing challenges of reframing our lifestyle assumptions and innovating sustainable technological and management approaches: to enable the airspace to continue supporting the needs of a burgeoning population currently subsisting on dwindling natural resources.

Air and the wider atmosphere have been substantially overlooked in terms of the benefits that they provide to society, as well as their vulnerability to a range of pressures. In common with other environmental media, historical policy measures – common and statute law, technological and market innovations, fiscal and other approaches to address pressures – have been fragmented, largely reacting on an issue-by-issue, and indeed substance-by-substance, basis only when particular gross impacts on human wellbeing have become evident. The consideration of wider systemic impacts on the ecosystem as a whole and of the implications of those impacts for human security, resulting from progressive and often insidious degradation of ecosystem services, has not significantly informed management of the air thus far. Indeed, this has been the case in many other spheres of environmental management and development planning. Unfortunately, unintended negative consequences from narrowly focused management responses remain commonplace.

Further research is required better to understand causal links across scales of space and time in ecosystem services, as well as between diverse stakeholder groups and their activities, and to determine effective responses and institutional arrangements to achieve this in a 'joined-up' way. The risks of continued failure to act systemically are becoming clearer, and the distribution of the benefits and costs of different responses also raises ethical issues. This is particularly the case for services provided by air, largely due to the transboundary, fluid and invisible nature of the medium, and because many impacts of degraded services manifest slowly and remotely in time and space from the pressures that generate them. The essentially ownerless status of air, and the extent to which its services have been overlooked, compounds these risks, as is reflected by their fragmented inclusion in law.

The substantial externalisation of most ecosystem services from the market creates further difficulties for bringing them into legal, policy and operational decision-making and action. The UK's *Stern Review of the Economics of Climate Change*[47] concluded that the unmet costs of climate change were the greatest market failure that the economist author had encountered. It is reasonable to conclude that many other services that have been historically overlooked or undervalued also represent significant market failures, making it all the more desirable to protect air as an integrated system through an Ecosystem Approach. Further research is required to develop reliable economic values, or value surrogates, to reflect the scale of benefits or costs associated with the many services provided by air and the wider atmosphere that are currently externalised from the market. The progressive inclusion of deduced values, imperfect but amenable to revision through an adaptive approach as further knowledge accumulates, will start to remedy current market failures. The same analysis of opportunities and gaps needs to be extended to the policy environment; this requires a bespoke study, with the services provided by air as the central integrating focus, rather than a continued emphasis on the disparate issue- and substance-specific control measures that framed legacy air quality management instruments.

A major factor revealed by effective, multiple outcome solutions is that, when considering the urban environment, there is a need to address in an integrated way all environmental media and all the facets of these deeply interconnected socio-economic systems. Solutions need to become interlinked, as otherwise isolated fixes to discretely identified 'problems' may inadvertently generate a wider range of negative outcomes for non-target issues.

7 | Rediscovering our place in the breathing space

So what, practically, do we need to do about insights drawn from a systemic perspective? This chapter is fundamentally about rediscovering our place in the common breathing space that supports the interests of us and of all life. It is about solutions that respect air and the atmosphere as a contiguous living system, and integrate them as key elements of sustainable planning and practice. It is only by safeguarding and rebuilding natural capacity, the processes and resilience of ecosystems, that we can expect to safeguard our own collective prospects for living fulfilled lives.

Important aspects of the paradigm shift

The arguments in this chapter rest on three fundamentally important aspects of the paradigm shift from largely retrospective and fragmented consideration of ecosystems towards a more connected, s ystemic approach to air and the atmosphere, and indeed to the wider biosphere and its human interactions. These aspects do not replace other sets of principles, such as those defining the Ecosystem Approach, but are useful touchstones to guide further shifts in thinking and practice. The first aspect is that nature has to be recognised correctly as a source of value. The second is that systemic solutions offer more sustainable outcomes. The third relates to empowering organisations to put these systemic perspectives into practice. These aspects are expanded below, before then turning to practical measures in a range of settings that will better develop a sustainable relationship with air and the atmosphere while continuing to meet our needs.

Nature as a source of value The first fundamentally important principle is that nature and its processes are among the most fundamental sources of value, realised through the multiple services that nature provides for human health, economic prospects and wider dimensions of wellbeing.

This assertion has been outlined many times already throughout this book in terms of the multiple, interlinked services that air and the atmosphere provide for humanity, many of them unrecognised in contemporary governance systems. However, it is at the very heart of the necessary societal paradigm shift towards sustainable development, since the trajectory of industrial society, as we have seen, has been one predicated on the concept of nature as a boundlessly exploitable resource. Consequently, market systems have focused on what we can make, build and trade from it, with regulatory and subsidy systems and engineering solutions generally implicitly reinforcing this world view. However, this narrow utilitarian model has many externalities, with a wide range of costs to ecosystems, ecosystem processes and services, and to the people, particularly future generations, who depend on them. We have seen how mined carbon, metals and nutrients end up pervading the atmosphere and wider biosphere, how the dispersion and remobilisation capacities of the air and other environmental media are exceeded by waste emissions across product life cycles – including, critically, beyond the end of those products' lives – and how the liquidation of critical habitats such as forests and wetlands destroys processes that regenerate the air system. Externalising the value of nature is a major market failure, undermining future resilience not only of the atmosphere and other ecosystems but of their capacities to continue to support humanity.

The evolution and logical forward trajectory of societal attitudes to nature conservation reflect our wider attitudes to our relationship with the ecosystems that support us. The journey of nature conservation has been one of initial recognition of dwindling species and habitats and the institution of protective measures external to the market. However, one consequence of this 'fortress conservation' approach, focusing on selected species and habitats, is that developers, local authority planners and others often perceive the protection of nature as a constraint on legitimate development. While protection of the rarest species and habitats remains vital if they are not to be lost, nature overall is losing the battle through declining connectivity and functioning across the wider landscape. It is towards this landscape scale that nature conservation has shifted its focus, looking to protect not just isolated patches of land and water but the freedom of species to move and interact and to do so in the places where farming, urban green spaces and other societal activities occur.[1] The agenda is moving towards an Ecosystem Approach, retaining a focus on the inherent value of nature but also recognising the multiple values that the natural world provides to society. This shift to framing nature as a source of value, not a constraint, is a vital step if the natural world is to be protected as an valued asset, rather than altruistically and in competition with the economy, as rising human demands impinge on dwindling natural resources.

Of course, recognising and then internalising the multiple values of air and the atmosphere across societal thinking and activities will challenge vested interests, planning approaches and fixed assumptions and practices founded on their extensive prior externalisation. However, it is not just the values that society derives directly from air and the atmosphere that need to be internalised. As we have seen, all environmental media are intimately interconnected, with the 'natural infrastructure' of ecosystems regenerating the structure and function of the atmosphere from which a wealth of direct and indirect societal benefits flow. So we also need to value ecosystems more generically – from the localised trees and reed beds that regulate microclimates and control pests right through to the great forests, oceans, rangelands and swamps that exhale oxygen to renew the ozone shield and otherwise regulate the global climate – for the many ways in which they contribute to the sustainability of our home 'bubble' of air.

Systemic solutions A second important principle is that management solutions should be systemic. This, too, is a significant shift in approach from our history of narrowly framed solutions to address equally narrowly defined problems. The environmental and financial costs of inputs to established environmental management technologies, and the unintended consequences arising from the limited consideration of their outputs, highlight the need for low-input solutions that optimise outcomes across multiple ecosystem services.

The concept of 'systemic solutions' is defined as 'low-input technologies using natural processes to optimise benefits across the spectrum of ecosystem services and their beneficiaries'.[2] Systemic solutions therefore contribute to sustainable development by averting unintended negative impacts and optimising benefits to all ecosystem service beneficiaries, increasing net economic value. Practical examples from the previous chapter include integrated constructed wetlands (ICWs) as multi-benefit solutions to integrated water and land management in preference to single-issue, input-intensive electromechanical wastewater treatment. The implementation of sustainable drainage systems (SuDS) and other low-input 'green infrastructure' methods to address linked air quality, green transport, aesthetic and amenity, hydrological, microclimate and biodiversity outcomes in urban areas is also a prime example of using ecosystem-based technologies that entail low management inputs yet yield a diverse and connected set of benefits that span historically disconnected policy areas.

Substantial obstacles to overcome in implementing 'systemic solutions' include legislation addressing issues in a fragmented way, associated ring-fenced budgets and established vested interests and management assumptions. One such significant and commonly encountered management assumption is that the 'land take' required for these systemic solutions is too costly, which may appear true if one thinks simply in terms of a single or a few beneficial outcomes. However, aggregated benefits and associated societal value rapidly accrue when one takes into account a wider suite of ecosystem services, yielding benefits for many societal interests, and the overall cumulative contribution to sustainability and 'liveability'. Farmers with ICWs on their landholdings who were surveyed in a study in Ireland certainly regarded these wetlands in this net value-generating light; they saw them as offering low-input waste management, improved landscape aesthetics, enhanced wildlife and fishing opportunities and other benefits, with wetland cells placed on natural drainage lines where rush vegetation and farm drainage had formerly been of minimal farming value.[3] Narrowly framed regulations and management budgets targeting single-issue outcomes pose significant barriers. However, increased flexibility in the implementation of legacy regulations, recognising their primary purpose rather than slavishly adhering to detailed subclauses, can prove an effective means to achieve greater overall public benefit through optimisation of outcomes across the spectrum of ecosystem services.[4]

Systemic solutions are not a panacea if applied merely as 'downstream' fixes. They are part of, and a means to accelerate, broader culture change towards more sustainable practice that necessarily entails connecting a wider network of interests in the formulation and design of such mutually beneficial solutions. Sectors that need to collaborate include spatial planners, engineers, regulators, managers, farming and other businesses, and researchers working on ways to quantify and optimise delivery of ecosystem services. This, of course, is entirely consistent with Ecosystem Approach principles, reflecting that ecosystem management is a matter of societal choice (Principle 1) and that it takes account of all forms of knowledge (Principle 11) and involves the participation of all relevant sectors of society (Principle 12).

Empowered planning We therefore have to make sure that the many institutions that society has established to govern and manage its activities are empowered to put these perspectives into practice. Do we then need to develop air and atmospheric plans at all scales? Or perhaps we need a meta-level 'ecosystems plan' to which all other plans are subservient? These ideas may seem rational in theory, but in practice are infeasible for two principal reasons. Firstly, the scales at which this top-down approach would have to operate range from the whole biosphere to the local community. Secondly, there will be contested ownership issues, as the owners of any single plan do not typically welcome invitations to become subservient to the owners of other plans. Anyway, this could conflict with the principle of devolved decision-making (addressed in, among other places, Ecosystem Approach Principles 2 and 7).

Furthermore, the problems we face today are rarely related to a lack of plans. In any locality in most developed nations, nests of overlapping plans address different areas of interest. A small subset of issues managed by these plans (with UK examples in brackets) includes air quality (Local Air Quality Management or LAQM), water resources (catchment management plans and river basin management plans of various descriptions), natural beauty (Area of Outstanding Natural Beauty plans), flood risk (catchment flood management plans), regional economic development (local enterprise partnerships), water abstraction (catchment abstraction management plans), national parks, tourism development, natural character, biodiversity, community cohesion (village design statements), green infrastructure, spatial development, nutrient control (nitrate vulnerable zones), low emission zones, and so on and so forth. We are not suffering from a deficit of plans, although perhaps there is a shortfall in their effective integration. So the idea of imposing a meta-plan for the atmosphere on top of this complexity is simply infeasible.

So the way we need to tackle the planning process itself needs to reflect the paradigm shift we seek. This entails moving from the hierarchical and narrowly disciplinary towards the systemic. In other words, the transformation of society will not occur by imposing new top-down mandates. What is necessary is to embed the Ecosystem Approach – or at the very least awareness of the ecosystem services framework for which the Ecosystem Approach offers broader geographical and socio-economic contexts – into these other multiple planning frameworks. This is empowering, if challenging, to plan-owners, planning authorities and local communities, as it exposes the wider ramifications of what they are seeking to achieve. And, by exposing these wider ramifications, people can then communicate in a common language, and innovate together to achieve outcomes that are more advantageous to a wider range of ecosystem service beneficiaries, or at least be more transparent about trade-offs where negative impacts on services and their associated beneficiaries are likely to occur.

This embedding of a systemic perspective is essential if people are to be empowered to make decisions that are optimally beneficial. It is true that this bottom-up approach does not explicitly favour the protection of air and the atmosphere, unlike a top-down mandate to include them in other plans. However, it is not only workable but more likely to succeed in practice in recognising the value of all ecosystems, including air, the atmosphere and the multiple ecosystem processes that regulate them, to all spheres of human interest.

What does this mean in practice?

In addressing the case studies of how we might do things differently in Chapter 6: *Thinking in a connected way*, we have already considered a number of ways in which air and atmospheric considerations might be embedded as part of a wider Ecosystem Approach. Some of these issues, and some further examples, are outlined here in a range of different settings.

Urban designs for life Various aspects of urban design have already been addressed. Factors such as the location and design of urban areas can have a significant impact on the air quality of cities. As discussed previously, the topographical nature and high populations of cities such as Mexico City, Santiago, Tehran, Los Angeles, Singapore and San Francisco render them particularly prone to smog, and the massive expansion of Chinese cities including Beijing and Shanghai is giving rise to serious smog and associated health and wider economic issues. Urban layout can work with or exacerbate natural air flows to overcome hazardous concentrations of air pollutants.

In addition, wider aspects of urban design, both in new developments and in redevelopment, can address air quality, climate change adaptation and mitigation, energy and transport efficiency, with implications for associated emission reductions. Design solutions can include, for example, measures such as green roofs, rain gardens (verges including green areas permitting water infiltration), open green spaces for people and green infrastructure in its widest sense, SuDS, street trees and urban forests, urban river restoration and related approaches that can make a significant difference to the retention and restoration of ecosystem services in urban environments, contributing to their overall sustainability and 'liveability'[5] as well as to people's health.[6] As simple a piece of natural infrastructure as an urban tree can have a major impact on a local scale. Indeed, the benefits of urban trees are becoming increasingly well known and quantified, including their capacity to sequester carbon. Urban trees also contribute to urban cooling, thereby reducing emissions from air conditioning, and to other energy savings; they promote urban biodiversity and cleanse the air, thereby improving health; and they make a positive contribution to public health improvements by softening hard urban streetscapes.[7]

Even such basic changes as improvements to building insulation can have major effects on the overall amount of climate-active, health-related and other emissions to the atmosphere. As far as building-related initiatives are concerned, a great deal of innovation is occurring in 'new builds', yet the sheer number of long-life buildings already in place demands a greater focus on refurbishing existing stock. This, of course, entails thinking not only about the 'fit' of buildings within the wider urban environment, but also about the role of trees and other vegetation in attenuating pollution through physical filtering effects and chemical transformation of pollutants as well as the wide range of additional green infrastructure benefits addressed above.

Value creation from the exploitation or emulation of nature Implicit in any consideration of the value of SuDS, river restoration and other green infrastructure in urban design, as well as other 'systemic solutions' such as ICWs and washlands in the wider landscape, is the fact that natural processes deliver value. This diversity of values has been almost entirely overlooked for so long, at least in the industrialised world, but is becoming increasingly recognised and positively exploited.

The redevelopment of Mayesbrook Park in east London, including the rehabilitation of the Mayes Brook, is an inspiring case study that illustrates how enhancement of a formerly undervalued inner-city habitat visibly delivers multiple societal benefits with associated economic value. Restoration of the Mayes Brook in Mayesbrook Park provided an opportunity to create an ecological and community focal point within a broader environmental regeneration project. The overall benefits arising from development shaped by ecosystem service considerations are substantial relative to the planned investment; the grand total of calculated lifetime benefits across the four ecosystem service categories (provisioning, regulatory, cultural and supporting) of around £27 million, including carbon sequestration by rehabilitated floodplains and tree growth, represents an excellent lifetime benefit-to-cost ratio of

7:1.[8] However, this economic value excludes a wide range of benefits assessed as likely to be significantly positive yet impossible to quantify with any confidence; such benefits include the regulation of air quality and the microclimate through tree planting and growth, with potentially substantial health benefits, as well as recreation and tourism gains through improved 'green space' facilities for local people and improvements in nutrient cycling and habitat for wildlife. The Mayesbrook Park case study concluded that ecosystem restoration was likely to provide a cost-effective means of achieving health benefits and social inclusion, a conclusion endorsed by a different study of ecosystem service benefits resulting from urban ecosystem restoration in Great Yarmouth in the east of England.[9]

The value of the regulatory ecosystem service of natural hazard protection is becoming increasingly recognised in coastal areas.[10] This is particularly the case for mangrove systems, which provide substantial cumulative value across a broad swathe of services, notably protection from natural hazards such as storm surges and tsunamis.[11] Indeed, 'markedly less' damage occurred in areas inland from the coast that were shielded from the impacts of the 2004 Indian Ocean tsunami by coastal tree vegetation.[12] A study of ecosystem services provided by the mangroves of the Indian megacity of Mumbai found that, although the mangroves have been significantly depleted by historical city development and continuing encroachment, they still provide substantial quantitative value to the built environment, people and future security of the city, particularly as a form of 'natural insurance' against changing climate, sea-level rise and other emerging sustainability challenges.[13] Indicative economic values for the protection of fresh water resources, food production, carbon sequestration, flood control, and industrial and domestic wastewater treatment were found to be significant. Ten other services that could not be monetised were assessed as 'significantly positive'. Most notable was the regulation of hazards and storms, including extreme events such as tsunamis, the impacts of which are likely to be disproportionately high given the low-lying profile of the city and its infrastructure and high real-estate values. Climate regulation was among the services considered to be 'significantly positive' in terms of both balances of greenhouse gases and the breaking down of the microclimate, and is of huge value in such a dense megacity as Mumbai. Cultural benefits were also found to be substantial. The protection of Mumbai's remaining mangroves, and ideally their restoration, may be critical for the future wellbeing of the city and its inhabitants due to their contribution to air, atmospheric and other ecosystem services. It therefore becomes critical for the future, linked vitality of the city and its 'natural infrastructure' that the value of these mangroves is included in planning on a systemic basis, rather than seeking to protect such important habitats retrospectively and based on an altruistic, preservation-based approach.

Another significant source of ecosystem services maintaining the viability of Mumbai, India's most populous city, is the Sanjay Gandhi National Park, a large protected area of 104 square kilometres (40 square miles) surrounded by the northern part of suburban Mumbai on three sides. Owing to its location, Sanjay Gandhi is one of the most visited national parks in the world. It also supports rich flora and fauna, and is a vital water resource for the city. However, Mumbai is facing serious and deteriorating air quality,[14] a problem that would be much worse were the massive 'green lung' of the national park not present within the city. It is certain that early deaths through air pollution and heat stress, already a substantial problem, would increase were the park not there, although the city's proximity to the coast is also important for the regeneration of its air. The principle of internalising the value of nature to safeguard air, atmospheric and other ecosystem services vital to human wellbeing is one from which we should learn. It is certainly one we should more routinely apply in how we plan urban development, protecting natural infrastructure and also re-establishing or emulating it through ecosystem process-centred 'green infrastructure' approaches. This provides a 'value added' means to bolster natural infrastructure and to de-risk urban areas.

Natural England, the natural environment regulator for England, undertook a place-based study of three upland regions to provide practical examples of how the Ecosystem Approach could be applied. Natural England worked with multiple local stakeholders in Bassenthwaite, the South Pennines and the South West Uplands to define land and water management and to address stakeholders' perceived needs in a cost-effective way.[15] Using mapping approaches and looking at ecosystem service supply, demand and opportunities for enhancement, Natural England and the stakeholders in this study identified land management options as a basis for the voluntary redirection of agri-environment and other funded land management approaches. These were linked to water company catchment management projects, and sought to optimise the net public value from investment across ecosystem services. The study found that the Ecosystem Approach, though not prescriptive about 'right answers', opened dialogue with stakeholders and increased community involvement in decision-making about the optimal ways of managing the landscapes they inhabit. Changing the emphasis from habitats and species to one of ecosystems was important, as it brought into consideration the many societal values that flow from ecological processes and interactions and shifted the focus towards issues of functionality (such as hydrology and soil science), which had not been prevalent in prior action planning, enhancing the resilience of landscapes. In the Keighley catchment area of the South Pennines, suggested improvements in catchment management arising from this engagement process would deliver a cost–benefit ratio of almost 1:3 over a 25-year period.

Lessons from the Natural England uplands pilot mirror those of Upstream Thinking[16] and of a more generic review of ecosystem service-based case studies:[17] that is, enhancement of ecosystem functioning and integrity tends to optimise benefits across the wide spectrum of ecosystem services. Putting the net benefits – for air quality, climate regulation and other atmospheric services, including the ways in which ecosystems contribute to them – at the centre of thinking represents a more sustainable basis for regional development. This approach to recognising and bringing into the mainstream the many services provided by landscapes is being pioneered through a process of stakeholder engagement across England under a government-sponsored programme of local nature partnerships.[18]

Retaining value in societal loops One of the oldest and most widely repeated definitions of pollution is 'a resource out of place', analogous to the botanical definition of a weed as 'a plant out of place'.[19] Some of these out-of-place resources are in fact valuable materials, and so permitting them to become pollutants makes no sense in terms of poisoning and other effects, and is also a waste of resources and sunk investment in a world in which greater resource efficiency is becoming increasingly essential. Consequently, measures to challenge the linear source–purify–manufacture–use–dispose product life that has dominated markets in the post-industrial world, moving towards a leaner and more cyclic economy and dependence on resources, are major contributors to protecting or improving the quality of air.

The mantra of reduce–reuse–recycle may be repeated often, but, like the definition of a 'resource out of place', it not only encodes useful principles but requires rather more urgent application across all spheres of societal activities. This can include not merely making further progress with waste resource recovery of all types, but also emerging measures such as landfill mining to reclaim historically disposed materials and to avoid pollutant releases into the air and other environmental media, including into biota.

An evolving policy environment When considering *empowered planning* earlier in this chapter, we found that the most strategic and acceptable approach to incorporating the value of services provided by air and the atmosphere lay not in 'special pleading' on an issue-by-issue basis, but in promoting the value added of the Ecosystem Approach across all policy areas. Embedding a systemic perspective is essential in order to empower people to make decisions of optimal benefit for ecosystem service outcomes, which are therefore equitable in addressing the needs of multiple stakeholders and of greatest cumulative economic value. It also provides a common conceptual framework, better to work across policy areas and planning perspectives.

The policy environment also needs a more flexible attitude to risk. Greater uncertainties are inherent in many 'systemic solutions' that work with natural processes in terms of delivering specific outcomes, such as the quantitative improvement of effluent quality that could be guaranteed by traditional electromechanical engineered solutions. However, accepting these risks is necessary to realise wider benefits arising from lower inputs and associated emissions, and from optimised delivery of multiple ecosystem services. To fail to promote systemic solutions in favour of more certain, if narrowly framed, outcomes is to reject their inherently greater sustainability and cumulative societal benefit per unit of input. This greater acceptance of risk has to be built into regulatory frameworks and regulator culture, steered by net societal outcomes against the ecosystem services framework.

The atmosphere is chaotic, in its scientific sense, so simple deterministic assumptions many not be a safe basis for wise policy development. We have to become more capable of understanding and managing uncertainty, since we understand only poorly the atmosphere and how it will react to interventions. However, we do know at least some of the processes – such as the role of vegetation in photosynthesis, the trapping of fine particulates and the breakdown of pollutants, the buffering of storm energy and water exchanges – that regenerate and regulate the air system, and we know that investment in these services may make a more valuable contribution to overall atmospheric health than 'downstream' technical fixes to problems caused by our activities and technologies. This, then, influences any deliberation of technology choice, as found, for example, in the balance of ecosystem-based versus engineering solutions to water resource management that depend on environmental factors, land manager engagement and societal priorities.[20] This approach is consistent with the Ecosystem Approach, decentralising decision-making to account for multiple value systems and balancing the conservation of ecosystem structure and functioning with the exploitation of those services.

A key conclusion emerging from analyses in this book is that it is necessary to integrate the values of air and the atmosphere across policy areas, rather than deal with them as special issues (alongside a wide range of competing special issues, such as nature and heritage conservation, water resource protection, and so on). One of the defining features of the transition needed in order to embed the value of air and the atmosphere in the mainstream of societal considerations is therefore that it should not be a matter of 'special pleading', but rather it should be considered as just part of the interconnected functional ecosystems that span air, water, earth and dependent human interests. Key to this is, of course, that ecosystems become a fundamental rather than a retrospective consideration in land use, industrial, trade, urban planning, transport and other considerations across all policy areas, safeguarding the many ways in which ecosystems underpin human interests.

However, there may be a role for addressing the air discretely, much as the global common of the world's oceans is addressed by the United Nations Convention on the Law of the Sea (UNCLOS). We have already considered calls for a global 'law of the atmosphere'[21] in Chapter 5: *Managing our impacts on air*. Substantial knowledge gaps need to be filled before such a law of the air could be formulated, including a valuation of the diverse services provided by the atmosphere.[22] However, knowledge gaps should not inhibit development of international agreements to the principle of balancing exploitation with conservation and the protection of wider interests in our common 'breathing space', a principle also enshrined for the sea in UNCLOS and for wider ecosystems by the Ecosystem Approach.

An evolving economy The evolving policy environment also depends on a novel economic environment that internalises the value of nature. This also has to be addressed on a systemic basis. A major obstruction to this at present is the 'ring-fenced' nature of budgets tied to specific, narrowly framed outcomes; this was illustrated in the previous chapter by the ways in which the management of wastewater treatment, air quality and climate change have been addressed to date.

There is also a role for new economic tools, such as payments for ecosystem services (PES) that integrate formerly overlooked ecosystem services into the economy. It was estimated that over 300 PES or PES-like schemes operated globally in 2010,[23] although the uptake of PES schemes around the world has since accelerated. In the UK, government promotes PES through commitments in its 2011 Natural Environment White Paper[24] and publication of a PES best practice guide.[25] PES schemes have been developed across the globe to address a range of biodiversity, water resource, flooding and other issues, including climate change and microclimate benefits. For example, the Costa Rica Payments for Environmental Services (Pagos por Servicios Ambientales or PSA) is a PES scheme that has been operational on the American continent since 1996. This scheme replaced an ineffective system of tax deductions with a national PES programme for services generated by forest and agro-forestry ecosystems. Participants entering this scheme are paid for four land use activities (protection of natural forest, establishment of timber plantations, natural forest regeneration, and establishment of agro-forestry systems) that produce a bundle of ecosystem services, including carbon sequestration, water quality protection, biodiversity protection and the provision of scenic beauty.[26] The scheme is funded by reallocating 3.5% of the revenue from a fossil fuel sales tax to the National Fund for Forestry Financing (Fondo Nacional de Financiamiento Forestal or FONAFIFO), worth about US$3.5 million per year. On top of this, the World Bank and other international aid donors also contribute.[27] In addition, the programme allows individual beneficiaries (hydroelectric plants, breweries, irrigated farms and other organisations benefiting from ecosystem services) to pay FONAFIFO, through which they may negotiate contracts with service providers. This Costa Rican scheme was formerly poorly targeted, with money allocated on a 'first come, first served',[28] uniform per-hectare basis, taking little account of local conditions and consequently making little contribution to deforestation.[29] A study has been undertaken to improve the efficiency of benefit production through spatial targeting and differentiation of payments.[30]

Another example of a PES scheme that includes climate change as part of forest protection operates in New Zealand, as touched upon in Chapter 6: *Thinking in a connected way*. While New Zealand undergoes a transition to a more urban-based economy, indigenous Māori landowners are expressing interest in markets for ecosystem services and in nature conservation payments to help maintain their livelihood and culture in North Island.[31] A Māori conservation reserve programme called Nga Whenua Rahui already provides a mechanism that enables landowners to allow land to remain in, or revert to, native bush. Development of wider markets for ecosystem services covering biodiversity protection, watershed restoration and carbon sequestration may be essential to secure resources to feed and offset the impacts of New Zealand's rapidly urbanising economy. Māori land is subject to a complex system of communal ownership, with decision-making authority for blocks of land vested in elected leaders. This creates some conflicts when there are opposing views about how land is best managed. Māori cultural values also mean that there is a different relationship between the people and their land than is found in Western economies. However, as we saw in Chapter 6, these complexities have the fortuitous consequence that environmental initiatives to prevent erosion, preserve water quality, restore forests and protect biodiversity resonate intuitively with *kaitiakitanga*, the Māori ethic of stewardship. Ngati Porou Whanui Forests Limited[32] has been established as a tribal cooperative bringing together Māori landowners and Māori agencies to benefit from market opportunities for ecosystem services, including both government incentives for the management of erodible land and foreign investment. Some forest areas may also be eligible for funding for carbon sequestration services. These new markets for ecosystem services may potentially become a significant element of economic growth among Māori people.

These are just two among many thousands of PES schemes operating around the world, with the REDD+ mechanism (addressed in Chapter 5: *Managing our impacts on air*) explicitly encouraging remuneration by industrialised nations for efforts by developing countries to retain carbon stores in situ in forests.

Another important economic consideration, also addressed by the Ecosystem Approach, is the need to think long term about the value of investment. At a national scale, this includes becoming more strategic about the targeting of subsidies, not frittering them away on short-term fixes – nor, indeed, regarding subsidies for strategically important technologies as expendable 'luxuries', as we are seeing all too commonly in post-'credit crunch' austerity measures in the 2010s. Only through long-term investment in the solutions needed for a more sustainable relationship between society and the atmosphere will sustained profit be achieved.

Remaining gaps There are, of course, substantial remaining gaps in the development of a truly sustainable relationship between humanity and the atmosphere and its many services. The periodic reports of the Intergovernmental Panel on Climate Change[33] are making progress in reviewing our collective knowledge about the workings of climatic systems and their intimate links with land use, forests, glaciers, mountains and other environmental systems, as well as the multiple human activities that modify them. However, there are many gaps remaining in our knowledge about the entire air and atmospheric system that warrant strategic exploration, particularly identification and valuation of the services that they provide. Valuation, in monetary terms but also recognising the many non-monetary values that atmospheric services bestow on humanity, is essential if the implications of those services for societal interests are to be grasped and incorporated through proportionate responses in governance systems.

Another 'elephant in the room' is, of course, the size and pace of human population growth. This is often overlooked but is at the heart of the scale of pressures we impose on the air and other environmental systems vital for our collective and continuing wellbeing.

Valuing the future

It would be impossible to overstate the intimacy with which people, nature and the air are interconnected. Indeed, we are inseparable; the combined processes of the natural world regenerate and maintain the atmosphere upon which all the elements of the natural world depend. As Lyall Watson stated in his insightful book *Supernature*:[34]

> there is a continuous communication not only between living things and their environment, but among all things living in that environment. An intricate web of interaction connects all life into one vast, self-maintaining system. Each part is related to every other part and we are all part of the whole, part of Supernature.

Under the paradigm from which we are emerging, air might have been thought of as an unlimited resource to mine and in which to dispose of waste, much like the rest of the natural world. We know differently now, even if our deep-rooted habits and vested interests are shifting only slowly towards sustainability in the face of so many adverse trends. We are now becoming increasingly aware of the many environmental and economic consequences of established lifestyles and inherited assumptions. This raises challenging equity issues, for both rich and poor worlds; for example, the low-lying islands threatened first by rising sea levels and other poor communities are those that have benefited least from historical human exploitation of the atmosphere and other natural resources. There are intergenerational equity issues too; the future will judge current generations, since we have in our hands both the opportunity and enough knowledge to choose a different pathway of development.

Revolution or evolution?

We are talking about a major paradigm shift rather than minor course correction. However, is this really revolution, or is it evolution? With hindsight, Europe's Industrial Revolution is seen as revolutionary, yet in reality it was a process of iterative, undirected change occurring most intensively over a period of two centuries. Likewise, the European Agricultural Revolution spanned three centuries of incremental innovation. Industrial and agricultural innovations continue today, and some countries are just embarking on their own such revolutions. Revolutions are generally evolutionary processes, each innovation progressing on the basis of prior inventions over a period of time dramatically compressed by hindsight.

As highlighted in Chapter 6: *Thinking in a connected way*, a similar 'telescope view' of the transition in the UK and much of the developed world throughout the twentieth century reveals that a great deal of progress has already been made in institutionalising a range of ecosystem services into various societal 'levers'. So an ecosystems revolution is already occurring, although many services provided by air and the atmosphere continue to be poorly recognised and addressed. We have already embarked on this journey in a haphazard sort of way. A far more directed, accelerated approach is necessary if we are to secure the scale of change required to underpin future human security and wellbeing.

There may be an assumption by those fearful of 'revolution' that all the old rules have to be discarded. Defensive behaviours commonly manifest themselves as a denial of the need to update a comfortable or habitual world view. However, it is in the nature of evolutionary change that future progress builds on where we are today. Therefore it should not be assumed automatically that a wholesale rejection and replacement of legislation and conceptual models is required. Our assessment reveals that we have made significant progress already in recognising and managing discrete health-related aspects of air pollution, and these remain important. However, we can add value to this baseline by factoring in a wider set of ecosystem services. This may create opportunities for us for synergistic solutions, addressing multiple benefits, and their associated economic benefits, and working with others whose actions may have co-benefits for our interests. Examples of this include considering air-related ecosystem service issues in 'non-air' management systems (for example, development planning, agricultural land use and wastewater management), considerations that are largely currently excluded. Many of these gaps could be addressed without the need for primary legislation, simply by issuing guidance requiring that legacy regulations should be interpreted using a framework of ecosystem services. This may include, for example, transparent assessment of the wider ramifications across the full suite of Millennium Ecosystem Assessment ecosystem services in established and accepted operational tools such as environmental impact assessments, strategic environmental assessments and 'programmes for measures' to achieve the goals of the EU Water Framework Directive. This aim of 'mainstreaming' the value of nature into all decision-making is indeed a key message of the UK's June 2011 Natural Environment White Paper, *The Natural Choice*,[35] albeit not explicit in its 92 commitments. Stronger and clearer government leadership is required to provide regulators and businesses with confidence and clear incentives to take this matter more seriously.

Where government is not able or willing to take the initiative, the common law may be applied to secure public rights in the airspace and in environmental media in general. This may require the development of legal actions initiated by civil society. Since the establishment of the Aarhus Convention,[36] which ensured access to environmental justice to signatory countries, such actions have become considerably cheaper.

To support legal, economic and policy development, as well as to provide a basis for internalising important atmospheric services into public and corporate decision-making, practical tools are required for operational use. These will be necessary to provide pragmatic means to help planners and other communities making on-the-ground decisions to internalise systemic thinking into those decisions. As indicated above, a great deal could be done by requiring assessments of the outcomes for all ecosystem services when implementing existing tools. Economic valuation can also help 'mainstream' systemic considerations by relating services, and impacts on service provision, to elements of societal wellbeing and the economic consequences of action or inaction. Indeed, there is already a significant track record of attempts to internalise the costs of degrading specific air-mediated services, for example in sulphur emissions trading to manage acid deposition[37] or emerging carbon trading schemes (such as the EU emissions trading system[38]), the UN Clean Development Mechanism[39] and agreements such as REDD+[40] that help address greenhouse gas emissions and hence the service of climate regulation on an economic basis. These tools will necessarily involve outreach to all affected stakeholders, requiring more participatory and deliberative techniques.[41]

Securing our shared 'breathing space'

Until now, air and the atmosphere have been substantially neglected environmental media, afflicted by their invisibility and fluidity in both physical and ownership senses. And, as for other environmental media, reactive and fragmented responses have tended to be instituted only as acute problems have become evident. Yet our common airspace is interdependent with all of life, including humanity and our diverse activities and interests, conferring vital and irreplaceable services on society.

Integration of the air and the many services that it provides into broader ecosystems assessment is long overdue. We need to become more considered, precautionary and respectful of our impacts upon it – and our direct and indirect impacts on all those who share it – in terms of the potential disruption of vital chemical, acoustic, light and other regimes. We need to be aware of the ramifications of our management of terrestrial and aquatic habitats, the functions of which maintain the structure and function of the atmosphere. This approach needs incrementally to moderate the legal, economic, equity, policy and other decision-making constructs, and to drive research along the lines of needs identified here to inform sustainable innovation and management.

There are no excuses now for ignorance about the vulnerabilities of the airspace and of all in society who share it. The only valid questions still to be answered are what we must do to safeguard it and when we will act, acknowledging the long-term consequences of our actions or, more insidious but no less informed, our inactions. As a matter of urgency we need progressively to modify our behaviour as if our very lives and livelihoods depended on it ... which frankly they do.

Perhaps what we need is a set of consensual, science-based but high-level 'atmospheric principles' that recognise important aspects of the services provided by air and the atmosphere. We already have such principles for the integrated management of water and land resources. A similar set of atmospheric principles might usefully guide the integration of this vital yet still commonly overlooked environmental medium into implementation of the Ecosystem Approach on scales ranging from the biospheric to the international and intranational, and through to local and corporate settings.

8 | Resolution for integrated management of the airspace

Despite its vast bulk, the fluidity, transboundary nature and lack of ownership of the airspace renders it not only the world's greatest 'common' but also the most commonly overlooked natural resource. An integrated approach to the recognition and wise use of this ecosystem is therefore long overdue, and needs to be instituted on a consistent international basis. This concluding chapter therefore posits its argument in the form of a proposed international resolution to safeguard the historically overlooked yet vital shared air and atmospheric system.

Building on prior successes

As we have seen, the Ecosystem Approach has, since its initial development in 1995, been defined as 'a strategy for the integrated management of land, water and living resources that promotes conservation and sustainable use in an equitable way'.[1] The Convention on Biological Diversity definition of the Ecosystem Approach goes on to state that: 'Essentially, it is a way of looking at and managing everything together. It recognizes that humans, with their cultural diversity, are an integral component of ecosystems.' Laudable and necessary as this extended definition is, air and the wider atmosphere are conspicuously missing. Consequently, as argued in Chapter 6: *Thinking in a connected way*, any sensible interpretation when implementing the Ecosystem Approach has to include air and the wider atmosphere together with their multiple services.

As yet, we lack the kind of consensual approach to managing the fluid common of the airspace that we have seen with integrated water and land management under integrated water resources management (IWRM). IWRM has become the dominant water management paradigm globally over the past several decades,[2] with international consensus substantially achieved under the four guiding principles of the Dublin Statement on Water and Sustainable Development, which was developed in the run-up to the Rio de Janeiro Earth Summit.[3] The four guiding Dublin principles are reproduced in Box 8.1.

Box 8.1 The four guiding Dublin principles for IWRM

Principle no. 1 Fresh water is a finite and vulnerable resource, essential to sustain life, development and the environment

Since water sustains life, effective management of water resources demands a holistic approach, linking social and economic development with protection of natural ecosystems. Effective management links land and water uses across the whole of a catchment area or groundwater aquifer.

Principle no. 2 Water development and management should be based on a participatory approach, involving users, planners and policy-makers at all levels

The participatory approach involves raising awareness of the importance of water among policy-makers and the general public. It means that decisions are taken at the lowest appropriate level, with full public consultation and involvement of users in the planning and implementation of water projects.

Principle no. 3 Women play a central part in the provision, management and safeguarding of water

Women play a pivotal role as providers and users of water and guardians of the living environment, requiring positive policies to address women's specific needs and to equip and empower them to participate at all levels in water resources programmes.

Principle no. 4 Water has an economic value in all its competing uses and should be recognized as an economic good

Within this principle, it is vital to recognize first the basic right of all human beings to have access to clean water and sanitation at an affordable price. Past failure to recognize the economic value of water has led to wasteful and environmentally damaging uses of the resource. Managing water as an economic good is an important way of achieving efficient and equitable use, and of encouraging conservation and protection of water resources.

A resolution for integrated management of the airspace

While the underlying tenet of Dublin principle no. 3 ('Women play a central part in the provision, management and safeguarding of water')

is important in the context of including all influential and vulnerable sectors in the management of natural resources, it is the other three principles that could be more directly redefined for the integrated consideration and co-management of the airspace. It is therefore suggested that they be reframed as the basis of an international resolution of 'atmospheric principles':

- *Atmospheric principle no. 1: The atmosphere is a finite and vulnerable resource, essential to sustain life, development and the environment* Since the atmosphere sustains life, effective management of atmospheric resources demands an holistic approach, linking social and economic development with protection of natural ecosystems. Effective management links air, land and water uses and the technological activities that influence them.
- *Atmospheric principle no. 2: The use and management of the airspace and human activities that influence it should be based on a participatory approach, involving users, planners and policy-makers at all levels* The participatory approach involves raising awareness of the importance of air and the atmosphere among policy-makers and the general public. It means that decisions are taken at the lowest appropriate level, with full public consultation and involvement of users in the planning and implementation of all human activities likely to have a significant impact on the properties and services of the atmosphere.
- *Atmospheric principle no. 3: Atmospheric services have economic values in all their competing uses and should be recognised as an economic good* Within this principle, it is vital to recognise first the basic right of all human beings to have access to fresh air and atmospheric protection. Past failure to recognise the economic value of atmospheric services, in both monetary and non-monetary terms, has led to environmental impacts that compromise the integrity and resilience of this vital resource. Managing the atmosphere as an economic good is an important way of achieving efficient and equitable use, and of encouraging mainstream consideration of the impacts of human activities upon this crucial resource.

One of the key strengths of this proposed approach to 'atmospheric principles' is that it builds on existing international agreements with respect to the integrated management of connected (water and land) resources. It explicitly demands that the value of the services provided by air and the atmosphere, and of impacts upon the atmospheric system, is brought into governance considerations.

It is therefore proposed that this set of 'atmospheric principles' be debated and subsequently adopted as a resolution by the international community as an important step towards the integrated assessment and valuation of air and the atmosphere – albeit arguably two decades too late, if one considers the genesis of the Ecosystem Approach and IWRM addressing the integrated management of water and land.

Supporting the resolution

Given that acceptance and the institutional infrastructure are already in place for implementation of both the Ecosystem Approach and the Dublin principles, uptake of the proposed 'atmospheric principles' should be relatively straightforward.

The proposed resolution is also more readily acceptable as it sets out the same kinds of broad principles as the other protocols, without specific objectives, measures or targets. It is therefore digestible by institutional structures in nation states across the world and by the companies that operate in and across their borders, delegating to them the freedom to enact, interact and report on a creative and flexible basis.

This supports the prior observations that protection of air and the atmosphere is best effected by embedding generic principles into devolved plans and decision-making forums. These principles should be based on the Ecosystem Approach but clarified by 'atmospheric principles' for the airspace, just as the Dublin principles clarify the Ecosystem Approach for land and water systems.

Annex: Ecosystem services and the Ecosystem Approach

The Millennium Ecosystem Assessment classification of ecosystem services

The concept of 'ecosystem services' arose around the late 1980s, initially as a pedagogic tool but also as a useful framework to guide international development. The Millennium Ecosystem Assessment defines ecosystem services as 'the benefits people obtain from ecosystems',[1] an anthropocentric approach to recognising the breadth of benefits that ecosystems provide to people. Diverse ecosystem service classification schemes have been developed since the late 1980s, generally addressing discrete habitat types within specific bioregions of the world, such as tropical wetlands, coral reefs, rainforests or temperate rangelands.

In order to compare the benefits provided by the world's major habitat types, the Millennium Ecosystem Assessment synthesised a harmonised international classification scheme of ecosystem services from this variety of pre-existing categorisations. Although neither perfect nor complete, the Millennium Ecosystem Assessment classification of ecosystem services is useful and a de facto global standard. Many users of the framework add additional services that are relevant to the environments in which they work. The Millennium Ecosystem Assessment classification, including some of these additional services, is set out in Table A.1 opposite.

The Ecosystem Approach

The Ecosystem Approach was advanced by the Convention on Biological Diversity (CBD) as 'a strategy for the integrated management of land, water and living resources that promotes conservation and sustainable use in an equitable way'.[2] First use of the term 'Ecosystem Approach' in a policy context occurred at the Earth Summit in Rio de Janeiro in 1992,[3] when it was adopted as a foundational concept of the CBD.[4] The Ecosystem Approach has since gained wider recognition and it is now an integral component of environmental policy, endorsed by the 2002 World Summit on Sustainable Development in Johannesburg,

TABLE A.1 Millennium Ecosystem Assessment classification of ecosystem services,[6] with commonly used addendum services in *italics*

Category	Services
Provisioning services	Fresh water Food (crops, fruit, fish, etc.) Fibre and fuel (timber, wool, etc.) Genetic resources (used for crop/stock breeding and biotechnology) Biochemicals, natural medicines, pharmaceuticals Ornamental resources (shells, flowers, etc.) Addenda to Millennium Ecosystem Assessment provisioning services *Natural energy harvesting (hydropower, wind, wave, tide, etc.)* *Aggregate and mineral harvesting*
Regulatory services	Air quality regulation Climate regulation (local temperature/precipitation, greenhouse gas sequestration, etc.) Water regulation (timing and scale of run-off, flooding, etc.) Natural hazard regulation (i.e. storm protection) Pest regulation Disease regulation Erosion regulation Water purification and waste treatment Pollination Addenda to Millennium Ecosystem Assessment regulatory services *Salinity regulation* (*soils*) *Fire regulation*
Cultural services	Cultural heritage Recreation and tourism Aesthetic value Spiritual and religious value Inspiration of art, folklore, architecture, etc. Social relations (fishing, grazing or cropping communities, etc.) Addenda to Millennium Ecosystem Assessment cultural services *Educational and research resources*
Supporting services	Soil formation Primary production Nutrient cycling Water recycling Photosynthesis (production of atmospheric oxygen) Provision of habitats

for example.[5] The Ecosystem Approach is implicit in the European Water Framework Directive.[7] It is also the recommended approach to halting the loss of biodiversity agreed in Gothenburg by the European Union Heads of Government and with regard to both natural and constructed wetlands by the Ramsar Convention.[8]

The CBD outlines that:

- Application of the ecosystem approach will help to reach a balance of the three objectives of the Convention. It is based on the application of appropriate scientific methodologies focused on levels of biological organization which encompass the essential processes, functions and interactions among organisms and their environment. It recognizes that humans, with their cultural diversity, are an integral component of ecosystems.

These three CBD principles are:

- the conservation of biological diversity;
- the sustainable use of its components; and
- the fair and equitable sharing of the benefits arising out of the utilization of genetic resources, including by appropriate access to genetic resources and by appropriate transfer of relevant technologies, taking into account all rights over those resources and to technologies, and by appropriate funding.[9]

The 12 principles of the Ecosystem Approach

The CBD provides 12 'complementary and interlinked' principles relating to implementation of the Ecosystem Approach.[10] These are reproduced in full below for the benefit of readers who may be unfamiliar or less familiar with them.

Principle 1: The objectives of management of land, water and living resources are a matter of societal choices Different sectors of society view ecosystems in terms of their own economic, cultural and society needs. Indigenous peoples and other local communities living on the land are important stakeholders and their rights and interests should be recognized. Both cultural and biological diversity are central components of the Ecosystem Approach, and management should take this into account. Societal choices should be expressed as clearly as possible. Ecosystems should be managed for their intrinsic values and for the tangible or intangible benefits for humans, in a fair and equitable way.

Principle 2: Management should be decentralized to the lowest appropriate level Decentralized systems may lead to greater efficiency, effectiveness and equity. Management should involve all stakeholders and balance local interests with the wider public interest. The closer management is to the ecosystem, the greater the responsibility, ownership, accountability, participation, and use of local knowledge.

Principle 3: Ecosystem managers should consider the effects (actual or potential) of their activities on adjacent and other ecosystems Management interventions in ecosystems often have unknown or unpredictable effects on other ecosystems; therefore, possible impacts need careful consideration and analysis. This may require new arrangements or ways of organization for institutions involved in decision-making to make, if necessary, appropriate compromises.

Principle 4: Recognizing potential gains from management, there is usually a need to understand and manage the ecosystem in an economic context Any such ecosystem-management programme should:

- reduce those market distortions that adversely affect biological diversity;
- align incentives to promote biodiversity conservation and sustainable use; and
- internalize costs and benefits in the given ecosystem to the extent feasible.

The greatest threat to biological diversity lies in its replacement by alternative systems of land use. This often arises through market distortions, which undervalue natural systems and populations and provide perverse incentives and subsidies to favour the conversion of land to less diverse systems.

Often those who benefit from conservation do not pay the costs associated with conservation and, similarly, those who generate environmental costs (e.g. pollution) escape responsibility. Alignment of incentives allows those who control the resource to benefit and ensures that those who generate environmental costs will pay.

Principle 5: Conservation of ecosystem structure and functioning, in order to maintain ecosystem services, should be a priority target of the Ecosystem Approach Ecosystem functioning and resilience depends on a dynamic relationship within species, among species and between species and their abiotic environment, as well as the physical and chemical interactions within the environment. The conservation and, where appropriate, restoration of these interactions and processes is of greater significance for the long-term maintenance of biological diversity than simply protection of species.

Principle 6: Ecosystems must be managed within the limits of their functioning In considering the likelihood or ease of attaining the management objectives, attention should be given to the environmental conditions that limit natural productivity, ecosystem structure, functioning and diversity. The limits to ecosystem functioning may be affected to different degrees by temporary, unpredictable or artificially maintained conditions and, accordingly, management should be appropriately cautious.

Principle 7: The Ecosystem Approach should be undertaken at the appropriate spatial and temporal scales The approach should be bounded by spatial and temporal scales that are appropriate to the objectives. Boundaries for management will be defined operationally by users, managers, scientists and indigenous and local peoples. Connectivity between areas should be promoted where necessary. The Ecosystem Approach is based on the hierarchical nature of biological diversity characterized by the interaction and integration of genes, species and ecosystems.

Principle 8: Recognizing the varying temporal scales and lag effects that characterize ecosystem processes, objectives for ecosystem management should be set for the long term Ecosystem processes are characterized by varying temporal scales and lag effects. This inherently conflicts with the tendency of humans to favour short-term gains and immediate benefits over future ones.

Principle 9: Management must recognize that change is inevitable Ecosystems change, including species composition and population abundance. Hence, management should adapt to the changes. Apart from their inherent dynamics of change, ecosystems are beset by a complex of uncertainties and potential 'surprises' in the human, biological and environmental realms. Traditional disturbance regimes may be important for ecosystem structure and functioning, and may need to be maintained or restored. The Ecosystem Approach must utilize adaptive management in order to anticipate and cater for such changes and events and should be cautious in making any decision that may foreclose options, but, at the same time, consider mitigating actions to cope with long-term changes such as climate change.

Principle 10: The Ecosystem Approach should seek the appropriate balance between, and integration of, conservation and use of biological diversity Biological diversity is critical both for its intrinsic value and because of the key role it plays in providing the ecosystem and other services upon which we all ultimately depend. There has been a tendency in the past to manage components of biological diversity either as protected or non-protected. There is a need for a shift to more flexible situations, where conservation and use are seen in context and the full range of measures is applied in a continuum from strictly protected to human-made ecosystems.

Principle 11: The Ecosystem Approach should consider all forms of relevant information, including scientific and indigenous and local knowledge, innovations and practices Information from all sources is critical to arriving at effective ecosystem management strategies. A much better knowledge of ecosystem functions and the impact of human use is desirable. All relevant information from any concerned area should be shared with all stakeholders and actors, taking into account, inter alia, any decision to be taken under Article 8(j) of the CBD. Assumptions behind proposed management decisions should be made explicit and checked against available knowledge and views of stakeholders.

Principle 12: The Ecosystem Approach should involve all relevant sectors of society and scientific disciplines Most problems of biological diversity management are complex, with many interactions, side effects and implications, and therefore should involve the necessary expertise and stakeholders at the local, national, regional and international level, as appropriate.

The five points of operational guidance of the Ecosystem Approach

In addition to the 12 principles outlined above, the CBD offers five points of operational guidance.[11] These are also reproduced in full below as a ready reference.

1. Focus on the relationships and processes within ecosystems The many components of biodiversity control the stores and flows of energy, water and nutrients within ecosystems, and provide resistance to major perturbations. A much better knowledge of ecosystem functions and structure, and the roles of the components of biological diversity in ecosystems, is required, especially to understand:

- ecosystem resilience and the effects to biodiversity loss (species and genetic levels) and habitat fragmentation;
- underlying causes of biodiversity loss; and
- determinants of local biological diversity in management decisions.

Functional biodiversity in ecosystems provides many goods and services of economic and social importance. While there is a need to accelerate efforts to gain new knowledge about functional biodiversity, ecosystem management has to be carried out even in the absence of such knowledge. The Ecosystem Approach can facilitate practical management by ecosystem managers (whether local communities or national policy-makers).

2. Enhance benefit-sharing Benefits that flow from the array of functions provided by biological diversity at the ecosystem level provide the basis of human environmental security and sustainability. The Ecosystem Approach seeks that the benefits derived from these functions are maintained or restored. In particular, these functions should benefit the stakeholders responsible for their production and management. This requires, inter alia: capacity-building, especially at the level of local communities managing biological diversity in ecosystems; the proper valuation of ecosystem goods and services; the removal of perverse incentives that devalue ecosystem goods and services; and, consistent with the provisions of the CBD, where appropriate, the replacement of perverse incentives with local incentives for good management practices.

3. Use adaptive management practices Ecosystem processes and functions are complex and variable. Their level of uncertainty is increased by their interaction with social constructs, which need to be better understood. Therefore, ecosystem management must involve a learning process, which helps to adapt methodologies and practices to the ways in which these systems are being managed and monitored. Implementation programmes should be designed to adjust to the unexpected, rather than to act on the basis of a belief in certainties. Ecosystem management needs to recognize the diversity of social and cultural factors affecting natural resource use. Similarly, there is a need for flexibility in policy-making and implementation. Long-term, inflexible decisions are likely to be inadequate or even destructive. Ecosystem management should be envisaged as a long-term experiment that builds on its results as it progresses. This 'learning by doing' will also serve as an important source of information to gain knowledge of how best to monitor the results of management and evaluate whether established goals are being attained. In this respect, it would be desirable to establish or strengthen the capacities of parties for monitoring.

4. Carry out management actions at the scale appropriate for the issue being addressed, with decentralization to the lowest level, as appropriate As noted in the description of the Ecosystem Approach, an ecosystem is a functioning unit that can operate at any scale, depending upon the problem or issue being addressed. This understanding should define the appropriate level for management decisions and actions. Often, this approach will imply decentralization to the level of local communities. Effective decentralization requires proper empowerment, which implies that the stakeholder has both the opportunity to assume responsibility and the capacity to carry out the appropriate action, and needs to be supported by enabling policy and legislative frameworks. Where common property resources are involved, the most appropriate scale for management decisions and actions would necessarily be large enough to encompass the effects of practices by all relevant stakeholders. Appropriate institutions would be required for such decision-making and, where necessary, for conflict resolution. Some problems and issues may require action at still higher levels, through, for example, transboundary cooperation, or even cooperation at global levels.

5. Ensure intersectoral cooperation As the primary framework of action to be taken under the CBD, the Ecosystem Approach should be fully taken into account in developing and reviewing national biodiversity strategies and action plans. There is also a need to integrate the Ecosystem Approach into agriculture, fisheries, forestry and other production systems that have an effect on biodiversity. Management of natural resources, according to the Ecosystem Approach, calls for increased intersectoral communication and cooperation at a range of levels (government ministries, management agencies, etc.). This might be promoted through, for example, the formation of inter-ministerial bodies within the government or the creation of networks for sharing information and experience.

Notes

1 The making of the atmosphere

1 www.maweb.org (accessed 22 December 2013).

2 uknea.unep-wcmc.org (accessed 22 December 2013).

3 Sigel et al. (2008); Weiner and Lowenstam (1989); Cuif et al. (2011).

4 Porter (2011); Cohen et al. (2011).

5 Koren et al. (2006).

6 Stuuta et al. (2009).

7 www.wrh.noaa.gov/twc/ (accessed 1 June 2007).

8 Zhu and Newell (1994).

9 Zhu and Newell (1998).

10 Fischetti (2012); Lavers et al. (2013).

11 EarthSky (2014).

12 See the US Office of Surface Water: water.usgs.gov/osw/ (accessed 20 June 2014).

13 www.worldwildlife.org/places/amazon (accessed 15 December 2013).

14 Harper et al. (2014).

15 www.maweb.org (accessed 22 December 2013).

2 Living in a bubble

1 As of 6 November 2013, according to Fédération Aéronautique Internationale criteria that define spaceflight as any flight to an altitude of over 100 kilometres (62 miles). See en.wikipedia.org/wiki/List_of_space_travelers_by_name (accessed 1 December 2013).

2 From Plato (2012).

3 Shapley (1967).

4 Shapley (1967).

5 Suzuki and McConnell (1997).

6 www.lsu.edu/deafness/HearingRange.html (accessed 1 December 2013).

7 Buck and Axel (1991).

8 Gilad et al. (2003).

9 Pantages and Dulac (2000).

10 Fountain (2006).

11 Small and Cohen (2004).

12 Thornes (1999).

13 Thornes (2008).

14 Lohrmann (1995); Drachmann (1961).

15 Gregory (2005).

16 De Decker (2009).

17 Global Wind Energy Council (2014).

18 For example, see the Sail Transport Network: www.sailtransportnetwork.org (accessed 7 Novemver 2014).

19 Vitousek et al. (1997).

20 Smith et al. (2004).

21 Erisman et al. (2008).

22 Sutton and van Grinsven (2011).

3 What does air do for us?

1 Millennium Ecosystem Assessment (2005a).

2 Millennium Ecosystem Assessment (2005b).

3 Millennium Ecosystem Assessment (2005a).

4 www.cbd.int/ecosystem/ (accessed 2 December 2013).

5 House et al. (2005).

6 Thornes et al. (2010).

7 Everard et al. (2013).

8 Walker (2007).

9 Thornes (2011).

10 Hardin (1968).

4 Abuses of the air

1 Plutchik and Kellerman (1980); Plutchik (2002).

2 BBC (2013a).

3 Bell et al. (2004).

4 Amann et al. (2005).

5 WHO (2010).

6 Moore (2014).

7 EC (2013).

8 Lancet (2014).

9 Roberts (2014).

10 Dixon et al. (1994).

11 www.wwf.org.uk/what_we_do/safeguarding_the_natural_world/forests/forest_work/amazon/amazon_ and_climate_change.cfm (accessed 15 December 2013).

12 Everard (2013).

13 Middleton and Goudie (2001).

14 Perry and Edgar (2014).

15 Mitra et al. (2005).

16 Millennium Ecosystem Assessment (2005b).

17 Mills and Harmens (2011).

18 Mills and Harmens (2011).

19 Hollaway et al. (2012).

20 US EPA (undated).

21 Farman et al. (1985).

22 Smith et al. (2004).

23 OECD (2013).

24 Willows and Hart (2013).

25 IPCC (2013).

26 WMO (2013).

27 WMO (2013).

28 The IPCC was established in 1988 by the WMO and UNEP, two UN organisations, to assess the scientific, technical and socio-economic information relevant to understanding the risk of human-induced climate change. It quantifies the magnitude of threat but also points to potential solutions for which bold political leadership will be required, and indeed a great deal of the content of this section on climate change is derived from various of the IPCC's authoritative reports. The work of the IPCC has been significant in collating global knowledge and data to drive emerging awareness and consensus about the magnitude of the threat posed by climate change and the need for substantial reductions in emissions if we are to avert its most extreme consequences. The IPCC has been an exemplar in international cooperation to address a common threat requiring international understanding and concerted action.

29 IPCC (1995).

30 IPCC (2007).

31 IPCC (2013).

32 IPCC (2014).

33 Wilford (2000).

34 Mann et al. (2009).

35 Cleetus (2013).

36 Brown (2003).

37 Stern (2006).

38 Radle (2007).

39 Krause (1993).

40 Dorrance et al. (1975).

41 Kavaler (1975); Calef et al. (1976); Knight (1984).

42 Votsi et al. (2012).

43 Floud et al. (2013).

44 Verheijen (1985).

45 Perry et al. (2008).

46 Longcore and Rich (2004).

47 Frank (1988).

48 Malakoff (2001).

49 Salmon (2003).

50 Brahic (2009).

51 Burks (1994); Baum et al. (1997); Pijnenburg et al. (1991); Knez (2001).

52 IARC (2007).

53 Schernhammer and Schulmeister (2004); Hansen (2001); Davis et al. (2001); Schernhammer et al. (2001); Bullough et al. (2006); Kloog et al. (2009).

54 Harrison (2007).

55 'No man is an island, Entire of itself, Every man is a piece of the continent, A part of the main.

If a clod be washed away by the sea, Europe is the less. As well as if a promontory were. As well as if a manor of thy friend's Or of thine own were: Any man's death diminishes me, Because I am involved in mankind, And therefore never send to know for whom the bell tolls; It tolls for thee' (John Donne, 1572–1631).

56 Crutzen and Stoermer (2000).

57 Crutzen (2002).

58 Samways (1999).

5 Managing our impacts on air

1 Costa (1997).

2 NSCA (2000).

3 Brimblecombe (1987).

4 Elcome (1999).

5 www.gov.uk/smoke-control-area-rules (accessed 8 November 2014).

6 HM Government (1956).

7 Longhurst et al. (2006).

8 HM Government (2000).

9 www.env.go.jp/en/coop/pollution.html (accessed 8 November 2013).

10 www.environment.gov.au/topics/environment-protection/air-quality/air-quality-standards (accessed 8 November 2013).

11 UNECE (1979).

12 National Academy of Sciences (1976).

13 Morrisette (1989).

14 ozone.unep.org/new_site/en/montreal_protocol.php (accessed 29 December 2013).

15 www.theozonehole.com/montreal.htm (accessed 14 November 2013).

16 Speth (2004: 95).

17 Newman et al. (2006).

18 EC (2008).

19 EU (2008).

20 www.air-quality.org.uk/20.php ((accessed 8 November 2014); Longhurst et al. (1996; 2009).

21 Department of the Environment (1997).

22 Beattie et al. (2001).

23 www.legislation.gov.uk/uksi/1997/3043/contents/made (accessed 16 November 2013).

24 Defra at al. (2007).

25 Defra (2009a; 2009b).

26 www.ipcc.ch (accessed 29 December 2013).

27 www.wmo.int (accessed 29 December 2013).

28 www.unep.org (accessed 29 December 2013).

29 IPCC (2006).

30 IPCC (2010).

31 CCTV (2013).

32 BBC (2007).

33 RCEP (2000).

34 Department for Energy and Climate Change (2008).

35 www.un-redd.org/AboutREDD/tabid/102614/Default.aspx (accessed 20 December 2013).

36 masdarcity.ae/en/ (accessed 21 June 2014).

37 Webster and Pagnamenta (2010).

38 Black (2007).

39 Razzouk (2013).

40 Stern (2006).

41 Everard (2011a).

42 Everard (2009).

43 www.darksky.org (accessed 20 December 2013).

44 BBC (2013b).

45 Walker (2007).

46 This definition is adapted from the one used by Natural England, the statutory nature conservation regulator in England (www.naturalengland.org.uk).

47 Everard et al. (2013).

48 Hardin (1968).

49 Scotland uses Roman land law, not common law.

50 Vogler (2001).

51 Gray (1991).

52 *William Aldred* [1610], as applied in *St Helen's Smelting Co. v Tipping* [1865] and many cases since.

53 Staddon et al. (2012).

54 Gray (1991).

55 Sax (1970).

56 'Land is not owned, it is holden, immediately or mediately, of the king' (Maitland 1886).

57 Everard and Moggridge (2012).

58 UNECE (1979).

59 Vogler (2001).

60 Najam (2001).

61 www.epa.gov/superfund/policy/cercla.htm (accessed 30 December 2013).

62 We may witness something similar following the BP Deepwater Horizon explosion of 2011. The Exxon settlement was about US$6 billion, whereas BP has already invested US$20 billion for reparation of damages.

63 The contingent valuation method is a form of stated preference valuation that uses information collected by means of surveys of people's preferences elicited through questionnaires. It is based on 'revealed preferences' in terms of how people act (which is then related to actual market data) rather than on the more nebulous 'expressed preferences' approach as applied in 'willingness to pay' studies.

64 www2.epa.gov/laws-regulations/summary-oil-pollution-act (accessed 29 December 2013).

65 Non-use or passive values include option value, bequest value and existence value.

66 Everard and Capper (2004); Everard and Appleby (2009).

67 HM Government (1956).

68 Defra (2007).

69 HM Government (2011).

70 OECD (2010); Smith et al. (2013).

71 www.valuing-nature.net (accessed 29 December 2013).

72 naturalcapitalproject.org/about.html (accessed 29 December 2013).

73 Castle and Nesary (1995).

74 For example, a right to expect water passed from your neighbour without significant diminution in quantity or quality, or, if not, freedom to seek reparations (the original 'polluter pays' principle) under riparian common law.

75 Yandle (2006).

76 Hayes et al. (2007).

77 Baldwin et al. (2008).

78 Baldwin et al. (2009).

79 Everard (2011a; 2013).

80 Ackerman (2009).

81 Smart et al. (2011).

6 Thinking in a connected way

1 Everard and Appleby (2009).

2 Everard (2011a); Everard et al. (2014).

3 Laffoley et al. (2004).

4 www.cbd.int/ecosystem.

5 www.cbd.int/ecosystem/principles.shtml.

6 Hayes et al. (2006).

7 Baldwin et al. (2008).

8 Baldwin et al. (2009).

9 Smart et al. (2011).

10 Sarkar and Agrawal (2010).

11 Bateman et al. (2001).

12 Stern (2006).

13 Rafaj et al. (2013).

14 Environment Agency (2005).

15 Nuttall et al. (1997).

16 Woods-Ballard et al. (2007).

17 Scholz et al. (2007); Harrington et al. (2011).

18 Everard et al. (2012).

19 Everard et al. (2014).

20 www.upstreamthinking.org (accessed 21 December 2013).

21 www.southwestwater.co.uk (accessed 21 December 2013).

22 www.wrt.org.uk (accessed 21 December 2013).

23 www.ofwat.gov.uk (accessed 29 December 2013).

24 www.gov.uk/catchment-sensitive-farming (accessed 21 December 2013).

25 Defra (2005).

26 www.gov.uk/government/publications/catchment-based-approach-improving-the-quality-of-our-water-environment (accessed 21 December 2013).

27 www.alfa-project.eu/en/about/index.php?mod=login&sel=setcookie (accessed 9 November 2014).

28 Perrot-Maître (2006).

29 Everard (2013).

30 Funk (2006).

31 Rowland (2012).

32 Robèrt (2002).

33 Feldman (2005).

34 Beckett et al. (2000).

35 www.treesforcities.org/ (accessed 16 November 2013); Harding (2009); Randall (2007); University of Washington (1998).

36 Longhurst et al. (2009).

37 For example, Moolgavkar and Luebeck (1996).

38 BusinessGreen.com (2010).

39 Umweltbundesamt (2010).

40 Santos et al. (2013).

41 Goodman et al. (2007).

42 www.bristol.gov.uk/page/environment/food-policy-bristol-and-food-charter (accessed 29 December 2013).

43 Directive 2009/28/EC of the European Parliament and of the Council of 23 April 2009 on the promotion of the use of energy from renewable sources and amending and subsequently repealing Directives 2001/77/EC and 2003/30/EC: eur-lex.europa.eu/legal-content/EN/ALL/?uri=CELEX:32009L0028 (accessed 19 October 2014).

44 www.greenpeace.org.uk/forests/palm-oil (accessed 21 D cember 2013).

45 Greenpeace (2008).

46 www.rspo.org (accessed 21 December 2013).

47 Stern (2006).

7 Rediscovering our place

1 Lawton (2010).

2 Everard and McInnes (2013).

3 Everard et al. (2012).

4 Everard (2011b).

5 Everard and Moggridge (2012).

6 Pugh et al. (2012).

7 www.treesforcities.org/about-us/information-resources/benefits-of-urban-trees/ (accessed 23 November 2013).

8 Everard et al. (2011).

9 Glaves et al. (2009).

10 Everard et al. (2010).

11 Voa et al. (2012).

12 Danielsen et al. (2005).

13 Everard et al. (2014).

14 Times of India (2013).

15 Waters et al. (2012).

16 www.upstreamthinking.org (accessed 22 December 2013).

17 Everard (2012).

18 www.gov.uk/government/policies/protecting-biodiversity-and-ecosystems-at-home-and-abroad/supporting-pages/local-nature-partnerships (accessed 29 December 2013).

19 For example, Royal Society of Chemistry (1996).

20 Everard (2013).

21 Najam (2001).

22 Thornes et al. (2010).

23 OECD (2010).

24 HM Government (2011).

25 Smith et al. (2013).

26 Wünscher et al. (2006).

27 OECD (2010).

28 FAO (2007).

29 Pfaff et al. (2007).

30 Wünscher et al. (2006).

31 Funk (2006).

32 www.ngatiporou.com (accessed 29 December 2013).

33 www.ipcc.ch (accessed 22 December 2013).

34 Watson (1974).

35 HM Government (2011).

36 UNECE (1998).

37 US EPA (2008).

38 EU (2010).

39 cdm.unfccc.int/ (accessed 16 November 2013).

40 Tropical Forest Group (2011).

41 Fish et al. (2011).

8 Management of the airspace

1 www.cbd.int/doc/meetings/cop/cop-09/media/cop9-press-kit-ea-en.pdf (accessed 23 December 2013).

2 Rahaman and Varis (2005).

3 www.wmo.int/pages/prog/hwrp/documents/english/icwedece.html (accessed 10 November 2014).

Annex 1

1 Millennium Ecosystem Assessment (2005a).

2 www.cbd.int/ecosystem/principles. shtml (accessed 24 November 2013).

3 Laffoley et al. (2004).

4 CBD (undated).

5 UN (2002).

6 Millennium Ecosystem Assessment (2005a).

7 EC (2000).

8 Ramsar Convention (2002); Beaumont et al. (2007).

9 www.cbd.int/convention/articles/default.shtml?a=cbd-01 (accessed 24 November 2013).

10 www.cbd.int/ecosystem/principles.shtml (accessed 24 November 2013).

11 www.cbd.int/ecosystem/operational.shtml (accessed 24 November 2013).

Bibliography

All URLs accessed 20 October 2014.

Ackerman, F. (2009) *Can We Afford the Future? Economics for a warming world*. London: Zed Books.

Amann, M., I. Bertok, J. Cofala, F. Gyarfas, C. Heyes, Z. Klimont, W. Schöpp and W. Winiwarter (2005) *Baseline Scenarios for the Clean Air for Europe (CAFE) Programme: Final Report. CAFE Scenario Analysis Report Nr. 1*. Laxenburg, Austria: International Institute for Applied Systems Analysis. ec.europa.eu/environment/archives/cafe/activities/pdf/cafe_scenario_report_1.pdf.

Baldwin, S. T., M. Everard, E. T. Hayes, J. W. S. Longhurst and J. R. Merefield (2008) 'Integrating local air quality and carbon management at a regional and local governance level: a case study of south west England'. In C. A. Brebbia and J. W. S. Longhurst (eds) *Air Pollution XVI*. Southampton: WIT Press, pp 159–68.

— (2009) 'Exploring barriers to and opportunities for the co-management of air quality and carbon in south west England: a review of progress. In C. A. Brebbia and V. Popov (eds) *Air Pollution XVII*. Southampton: WIT Press.

Bateman, I., B. Day, I. Lake and A. Lovett (2001) *The Effect of Road Traffic on Residential Property Values: A literature review and hedonic pricing study*. Edinburgh: Scottish Executive Development Department.

Baum, A., R. West, J. Weinman, S. Newman and C. McManus (eds) (1997) *Cambridge Handbook of Psychology, Health and Medicine*. Cambridge: Cambridge University Press.

BBC (2007) 'Climate bill's 60% emission cut'. BBC News, 6 November. news.bbc.co.uk/1/hi/uk_politics/7080580.stm.

— (2013a) '10 dangerous things in Victorian/Edwardian homes'. *BBC News Magazine*, 16 December. www.bbc.co.uk/news/uk-25259505.

— (2013b) 'Dark-sky status awarded to Northumberland Park area'. BBC News, Tyne and Wear, 9 December. www.bbc.co.uk/news/uk-england-tyne-25260186.

Beattie, C. I., J. W. S. Longhurst and N. K. Woodfield (2001) 'Air quality management: evolution of policy and practice in the UK as exemplified by the experience of English local government'. *Atmospheric Environment* 35(8): 1479–90.

Beaumont, N. J., M. C. Austen, J. P Atkins, D. Burdon, S. Degraer, T. P. Dentinho, S. Derous, P. Holm, T. Horton and E. van Ierland (2007) 'Identification, definition and quantification of goods and services provided by marine biodiversity: implications for the ecosystem approach'. *Marine Pollution Bulletin* 54: 253–65.

Beckett, K. P., P. Freer-Smith and G. Taylor (2000) 'Effective tree species for local air-quality management'. *Journal of Arboriculture* 26(1): 12–19.

Bell, M. L., D. L. Davis and T. Fletcher (2004) 'A retrospective assessment of mortality from the London smog episode of 1959: the role of influenza and pollution'. *Environmental Health Perspectives* 112(1): 6–8. doi: 10.1289/ehp.6539.

Black, R. (2007) 'Humans blamed for climate change'. BBC News, 2 February. news.bbc.co.uk/2/hi/science/nature/6321351.stm.

Brahic, C. (2009) 'Wildlife confused by polarised light pollution'. *New Scientist*, 8 January. www.newscientist.com/article/dn16380-wildlife-confused-by-polarised-light-pollution.html#.Ur8jAtJdWSo.

Brimblecombe, P. (1987) *The Big Smoke: A history of air pollution in London since medieval times*. York: Methuen.

Brown, L. R. (2003) *Eco-Economy: Building an economy for the Earth*. London: Earthscan.

Buck, L. and R. Axel (1991) 'A novel multigene family may encode odorant receptors: a molecular basis for odor recognition'. *Cell* 65(1): 175–87. doi: 10.1016/0092-8674(91)90418-X.

Bullough, J. D., M. S. Rea and M. G. Figueiro (2006) 'Of mice and women: light as a circadian stimulus in breast cancer research'. *Cancer Causes and Control* 17(4): 375–83.

Burks, S. L. (1994) *Managing Your Migraine*. Totowa, NJ: Humana Press.

BusinessGreen.com (2010) 'Germany sets out zero-carbon road map'. BusinessGreen.com, 8 July. www.businessgreen.com/bg/news/1802455/germany-sets-zero-carbon-road-map.

Calef, G., E. A. DeBock and G. M. Lortie (1976) 'The reaction of barren-ground caribou to aircraft'. *Arctic* 29(4): 210–12.

Carson, R. (1962) *Silent Spring*. Boston, MA: Houghton Mifflin.

Castle, E. N. and M. R. Nesary (1995) 'Putting a price tag on nature: problems and techniques'. *Montana Business Quarterly*, 22 June.

CBD (undated) 'Operational guidance for application of the ecosystem approach'. Convention on Biological Diversity (CBD) website. www.cbd.int/ecosystem/operational.shtml.

CCTV (2013) 'China: rich countries should live up to commitments on climate change'. CCTV, 14 November. english.cntv.cn/program/newsupdate/20131114/103907.shtml.

Cleetus, R. (2013) *Overwhelming Risk: Rethinking flood insurance in a world of rising seas*. Cambridge, MA: Union of Concerned Scientists. www.ucsusa.org/assets/documents/global_warming/Overwhelming-Risk-Full-Report.pdf.

Cohen, P. A., J. W. Schopf, N. J. Butterfield, A. B. Kudryavtsev and F. A. MacDonald (2011) 'Phosphate biomineralization in mid-Neoproterozoic protists'. *Geology* 9(6): 539–42. doi: 10.1130/G31833.1.

Costa, C. D. N. (1997) *Dialogues and Letters: Seneca*. London: Penguin Books.

Crutzen, P. J. (2002) 'Geology of mankind'. *Nature* 415(6867): 23.

Crutzen, P. J. and E. F. Stoermer (2000) 'The "Anthropocene"'. *Global Change Newsletter* 41: 17–18.

Cuif, J.-P., Y. Dauphin and J. E. Sorauf

(2011) *Biominerals and Fossils Through Time*. Cambridge: Cambridge University Press.

Danielsen, F., M. K. Sørensen, M. F. Olwig, V. Selvam, F. Parish, N. D. Burgess, T. Hiraishi, V. M. Karunagaran, M. S. Rasmussen, L. B. Hansen, A. Quarto and N. Suryadiputra (2005) 'The Asian tsunami: a protective role for coastal vegetation'. *Science* 310: 643.

Davis, S., D. K. Mirick and R. G. Stevens (2001) 'Night shift work, light at night, and risk of breast cancer'. *Journal of the National Cancer Institute* 93(20): 1557–62.

De Decker, K. (2009) 'Wind powered factories: history (and future) of industrial windmills'. *Low-tech Magazine*, 8 October. www.lowtechmagazine.com/2009/10/history-of-industrial-windmills.html.

Defra (2005) *Making Space for Water: Taking forward a new government strategy for flood and coastal erosion risk management in England. First government response to the autumn 2004 Making Space for Water consultation exercise, March 2005*. London: Department for Environment, Food and Rural Affairs (Defra).

— (2007) *An Introductory Guide to Valuing Ecosystem Services*. London: Department for Environment, Food and Rural Affairs (Defra).

— (2009a) *Local Air Quality Management: Policy guidance PG(09)*. London: Department for Environment, Food and Rural Affairs (Defra). archive.defra.gov.uk/environment/quality/air/air quality/ local/guidance/documents/laqm-policy-guidance-part4.pdf.

— (2009b) Local Air Quality Management: Technical guidance LAQM. TG(09). London: Department for Environment, Food and Rural Affairs (Defra). www.gov.uk/government/publications/local-air-quality-management-technical-guidance-laqm-tg-09.

Defra, Scottish Government, WAG and DoE/NI (2007) *The Air Quality Strategy for England, Scotland, Wales and Northern Ireland*. London: Department for Environment, Food and Rural Affairs (Defra) in partnership with the Scottish Government, Welsh Assembly Government (WAG) and Department of the Environment Northern Ireland (DoE/NI).

Department for Energy and Climate Change (2008) 'Climate Change Act 2008'. London: HM Government. www.legislation.gov.uk/ukpga/2008/27/contents.

Department of the Environment (1997) *The United Kingdom National Air Quality Strategy*. London: Department of the Environment.

Dixon, R. K., S. Brown, R. A. Houghton, A. M. Solomon, M. C. Trexler and J. Wisniewski (1994) 'Carbon pools and flux of global forest ecosystems'. *Science* 263: 185–91.

Dorrance, M., P. J. Savage and D. E. Huff (1975) 'Effects of snowmobiles on white-tailed deer'. *Journal of Wildlife Management* 39(3): 563–9.

Drachmann, A. G. (1961) 'Heron's windmill'. *Centaurus* 7: 145–51.

EarthSky (2014) 'How much do oceans add to world's oxygen?' *EarthSky*, 8 June. earthsky.org/earth/how-much-do-oceans-add-to-worlds-oxygen.

EC (2000) 'Directive 2000/60/EC of the European Parliament and of

the Council establishing a framework for the community action in the field of water policy'. Brussels: European Commission (EC).
— (2008) 'Framework Directive 96/62/EC on Ambient Air Quality Assessment and Management'. Brussels: European Commission (EC). ec.europa.eu/environment/air/quality/legislation/existing_leg.htm.
— (2013) 'The Clean Air Policy Package'. Brussels: European Commission (EC). ec.europa.eu/environment/air/clean_air_policy.htm.
Elcome, D. (1999) *The Fragile Environment: Pollution and abuse*. Cheltenham: Nelson Thornes.
Environment Agency (2005) *A Better Place? State of the environment 2005*. Bristol: Environment Agency.
Erisman, J. W., M. A. Sutton, J. Galloway, Z. Klimont and W. Winiwarter (2008) 'How a century of ammonia synthesis changed the world'. *Nature Geoscience* 1(10): 636. doi: 10.1038/ngeo325.
EU (2008) *Directive 2008/50/EC of the European Parliament and of the Council of 21 May 2008 on Ambient Air Quality and Cleaner Air for Europe*. Brussels: European Union (EU). eur-lex.europa.eu/LexUriServ/LexUriServ.do?uri=OJ:L:2008:152:0001:0044:EN:PDF.
— (2010) 'EU Emissions Trading System (EU ETS)'. Brussels: European Union (EU). ec.europa.eu/clima/policies/ets/index_en.htm.
Everard, M. (2009) *The Business of Biodiversity*. Ashurst: WIT Press.
— (2011a) *Common Ground: The sharing of land and landscapes for sustainability*. London: Zed Books.
— (2011b) 'Why does "good ecological status" matter? *Water and Environment Journal* 26(2): 165–74. doi: 10.1111/j.1747-6593.2011.00273.x.
— (2012) 'What have rivers ever done for us? Ecosystem services and river systems'. In P. J. Boon and P. J. Raven (eds) *River Conservation and Management*. Chichester: John Wiley, pp. 313–24.
— (2013) *The Hydropolitics of Dams: Engineering or ecosystems?* London: Zed Books.
Everard, M. and T. Appleby (2009) 'Safeguarding the societal value of land'. *Environmental Law and Management* 21: 16–23.
Everard, M. and K. Capper (2004) 'Common Law and River Conservation: The Case for Whole System Thinking'. *Environmental Law and Management* 16: 135–44.
Everard, M. and R. McInnes (2013) 'Systemic solutions for multi-benefit water and environmental management'. *Science of the Total Environment* 461–62: 170–9.
Everard, M. and H. L. Moggridge (2012) 'Rediscovering the value of urban rivers'. *Urban Ecosystems* 15(2): 293–314. doi: 10.1007/s11252-011-0174-7.
Everard, M., R. Harrington and R. J. McInnes (2012) 'Facilitating implementation of landscape-scale integrated water management: the integrated constructed wetland concept'. *Ecosystem Services* 2: 27–37. doi: 10.1016/j.ecoser.2012.08.001.
Everard, M., R. R. S. Jha and S. Russell (2014) 'The benefits of fringing mangrove systems to Mumbai'. *Aquatic Conservation* 24(2): 256–74.
Everard, M., L. Jones and B. Watts (2010) 'Have we neglected the

societal importance of sand dunes? An ecosystem services perspective'. *Aquatic Conservation: Marine and Freshwater Ecosystems* 20: 476–87.

Everard, M., L. Shuker and A. Gurnell (2011) *The Mayes Brook Restoration in Mayesbrook Park, East London: An ecosystem services assessment*. Bristol: Environment Agency. www.theriverstrust.org/projects/water/Mayes%20brook%20restoration.pdf.

Everard, M., B. Pontin, T. Appleby, C. Staddon, E. T. Hayes, J. Barnes and J. Longhurst (2013) 'Air as a common good'. *Environmental Science and Policy* 33: 354–68.

Everard, M., J. Dick, H. Kendall, R. I. Smith, R. W. Slee, L. Couldrick, M. Scott and C. McDonald (2014) 'Improving coherence of ecosystem service provision between scales'. *Ecosystem Services* 9: 66–74. doi: 10.1016/j.ecoser.2014.04.006.

FAO (2007) *The State of Food and Agriculture 2007: Paying farmers for environmental services*. Rome: Food and Agricultural Organization of the United Nations (FAO). ftp://ftp.fao.org/docrep/fao/010/a1200e/a1200e00.pdf.

Farman, J. C., B. G. Gardiner and J. D. Shanklin (1985) 'Large losses of total ozone in Antarctica reveal seasonal ClOx/NOx interaction'. *Nature* 315(6016): 207. doi: 10.1038/315207a0.

Feldman, A. M. and R. Serrano (2005) *Welfare Economics and Social Choice Theory*. New York, NY: Springer.

Fischetti, M. (2012) 'Mysterious atmospheric river soaks California, where megaflood may be overdue'. *Scientific American*, 30 November. blogs.scientificamerican.com/observations/2012/11/30/mysterious-atmospheric-river/.

Fish, R., J. Burgess, J. Chilvers, A. Footitt, R. Haines-Young, D. Russel, R. K. Turner and D. M. Winter (2011) *Participatory and Deliberative Techniques for Embedding an Ecosystems Approach into Decision Making: An introductory guide*. London: Department for Environment, Food and Rural Affairs.

Floud, S., M. Blangiardo, C. Clark et al. (2013) 'Exposure to aircraft and road traffic noise and associations with heart disease and stroke in six European countries: a cross-sectional study'. *Environmental Health* 12: 89. doi: 10.1186/1476-069X-12-89. www.ehjournal.net/content/12/1/89.

Fountain, H. (2006) 'This plant has the sense of smell (loves tomatoes, hates wheat)'. *New York Times*, 3 October.

Frank, K. D. (1988) 'Impact of outdoor lighting on moths'. *Journal of the Lepidopterists' Society* 42: 63–93.

Funk, J. (2006) 'Maori farmers look to environmental markets in New Zealand'. *Ecosystem Marketplace*, 24 January. www.ecosystemmarketplace.com/pages/dynamic/article.page.php?page_id=4097§ion=home&eod=1.

Gilad, Y., O. Man, S. Pääbo and D. Lancet (2003) 'Human specific loss of olfactory receptor genes'. *Proceedings of the National Academy of Sciences of the United States of America* 100(6): 3324–7. doi: 10.1073/pnas.0535697100.

Glaves, P., D. Egan, K. Harrison and R. Robinson (2009) *Valuing Ecosystem Services in the East of England*. Cambridge: East of England Environment Forum, East of England Regional Assembly and Government Office East England.

Global Wind Energy Council (2014) *Global Wind Statistics 2013*. Brussels: Global Wind Energy Council. www.gwec.net/wp-content/uploads/2014/02/GWEC-PRstats-2013_EN.pdf.

Goodman, J., M. Laube and J. Schwenk (2007) 'Curitiba's bus system is model for rapid transit'. *Urban Habitat*. urbanhabitat.org/files/25.Curitiba.pdf.

Gray, K. (1991) 'Property in thin air'. *Cambridge Law Journal* 50(2): 252–307.

Greenpeace (2008) *The Hidden Carbon Liability of Indonesian Palm Oil*. Amsterdam: Greenpeace International. www.greenpeace.org.uk/files/pdfs/forests/hidden-carbon-liability-of-palm-oil.pdf.

Gregory, R. (2005) *The Industrial Windmill in Britain*. Chichester: Phillimore.

Hansen, J. (2001) 'Increased breast cancer risk among women who work predominantly at night'. *Epidemiology* 12(1): 74–7.

Hardin, G. (1968) 'The tragedy of the commons'. *Science* 162(3859): 1243–8.

Harding, A. (2009) 'Tree-lined streets may cut city kids' asthma risk'. *Reuters Health*, 9 May. www.reuters.com/article/healthNews/idUSKEN96671020080509.

Harper, A. B., I. T. Baker, A. S. Denning, D. A. Randall, D. Dazlich and M. Branson (2014) 'Impact of evapotranspiration on dry season climate in the Amazon forest'. *Journal of Climate* 27(2): 574–91. doi: 10.1175/JCLI-D-13-00074.1.

Harrington, R., P. Carroll, S. Cook, C. Harrington, M. Scholz and R. J. McInnes (2011) 'Integrated constructed wetlands: water management as a land-use issue, implementing the "Ecosystem Approach"'. *Water Science and Technology* 63(12): 2929–37. doi: 10.2166/wst.2011.591.

Harrison, E. Z. (2007) 'Health impacts of composting air emissions'. *BioCycle* November: 44–50.

Hayes, E. T., T. J. Chatterton, N. S. Leksmono and J. W. S. Longhurst (2006) 'Integrating climate change management into the local air quality management process at a local and regional governance level in the UK'. In J. W. S. Longhurst and C. A. Brebbia (eds) *Air Pollution XIV*. Southampton: WIT Press, pp. 439–46.

Hayes, E. T, N. S. Leksmono, T. J. Chatterton, J. K. Symons, S. T. Baldwin and J. W. S. Longhurst (2007) 'Co-management of carbon dioxide and local air quality pollutants: identifying the "win–win" actions'. In *Proceedings of the 14th International Union of Air Pollution Prevention and Environmental Protection Associations (IUAPPA) World Congress 2007*. Brisbane, 9–13 September.

HM Government (1956) 'Clean Air Act 1956'. www.legislation.gov.uk/ukpga/Eliz2/4-5/52/contents/enacted.

— (2000) 'The Pollution Prevention and Control (England and Wales) Regulations 2000'. www.legislation.gov.uk/uksi/2000/1973/contents/made.

— (2011) *The Natural Choice: Securing the value of nature*. London: The Stationery Office.

Hollaway, M. J., S. R. Arnold, A. J. Challinor and L. D. Emberson (2012) 'Intercontinental trans-boundary contributions to

ozone-induced crop yield losses in the northern Hemisphere'. *Biogeosciences* 9: 271–92.
House, J., V. Brovkin, R. Betts, R. Constanza, M. A. S. Dias, B. Holland, C. Le Quéré, N. K. Phat, U. Riebesell and M. Scholes (2005) 'Climate and air quality'. In Millennium Ecosystem Assessment, *Ecosystems and Human Well-being: Current state and trends*. Washington, DC: Island Press. www.unep.org/maweb/en/Condition.aspx.
IARC (2007) 'IARC Monographs Programme finds cancer hazards associated with shiftwork, painting and firefighting'. Press release no. 180, 5 December. Lyon: International Agency for Research on Cancer (IARC). www.iarc.fr/en/media-centre/pr/2007/pr180.html.
IPCC (1995) *IPCC Second Assessment: Climate change 1995*. Geneva: Intergovernmental Panel on Climate Change (IPCC). www.ipcc.ch/pdf/climate-changes-1995/ipcc-2nd-assessment/2nd-assessment-en.pdf.
— (2006) 'Principles governing IPCC work'. Geneva: Intergovernmental Panel on Climate Change (IPCC). www.ipcc.ch/pdf/ipcc-principles/ipcc-principles.pdf.
— (2007) *Climate Change 2007: Synthesis report*. Geneva: Intergovernmental Panel on Climate Change (IPCC). www.ipcc.ch/publications_and_data/ar4/syr/en/contents.html.
— (2010) *Understanding Climate Change: 22 years of IPCC assessment*. Geneva: Intergovernmental Panel on Climate Change (IPCC). www.ipcc.ch/pdf/press/ipcc_leaflets_2010/ipcc-brochure_understanding.pdf.
— (2013) *Climate Change 2013: The physical science basis*. Geneva: Intergovernmental Panel on Climate Change (IPCC). www.ipcc.ch/report/ar5/wg1/#.Ur7_PtJdWSo.
— (2014) *Climate Change 2014: Impacts, adaptation, and vulnerabiity – summary for policymakers*. Geneva: Intergovernmental Panel on Climate Change (IPCC). ipcc-wg2.gov/AR5/images/uploads/IPCC_WG2AR5_SPM_Approved.pdf.
Kavaler, L. (1975) *Noise: The new menace*. New York, NY: John Day Company.
Kloog, I., A. Haim, R. G. Stevens and B. A. Portnov (2009) 'Global co-distribution of light at night (LAN) and cancers of prostate, colon, and lung in men'. *Chronobiology International* 26(1): 108–25.
Knez, I. (2001) 'Effects of colour of light on nonvisual psychological processes'. *Journal of Environmental Psychology* 21(2): 201.
Knight, R. (1984) 'Responses of wintering bald eagles to boating activity'. *Journal of Wildlife Management* 48(3): 999–1004.
Koren, I., Y. J. Kaufman, R. Washington, M. C. Todd, Y. Rudich, J. V. Martins and D. Rosenfeld (2006) 'The Bodélé depression: a single spot in the Sahara that provides most of the mineral dust to the Amazon rainforest'. *Environmental Research Letters* 1 (October–December).
Krause, B. (1993) 'The niche hypothesis'. *The Soundscape Newsletter*, 6 June.
Laffoley, D., E. Maltby, M. A. Vincent, L. Mee, E. Dunn, P. Gilliland, J. P. Hamer, D. Mortimer and D. Pound (2004) *The Ecosystem Approach: Coherent actions for marine and coastal environments.*

A report to the UK government. Peterborough: English Nature.

Lancet (2014) 'Clearing the air over Europe, and elsewhere'. *The Lancet* 383(9911): 1. www.thelancet.com/journals/lancet/article/PIIS0140-6736(13)62714-2/fulltext?elsca1=ETOC-LANCET&elsca2=email&elsca3=E24A35F.

Lavers, D. A., R. P. Allan, G. Villarini, B. Lloyd-Hughes, D. J. Brayshaw and A. J. Wade (2013) 'Future changes in atmospheric rivers and their implications for winter flooding in Britain'. *Environmental Research Letters* 8(3): 034010. iopscience.iop.org/1748-9326/8/3/034010/article.

Lawton, J. (2010) *Making Space for Nature: A review of England's wildlife sites and ecological network.* London: Department for Environment, Food and Rural Affairs. archive.defra.gov.uk/environment/biodiversity/documents/201009space-for-nature.pdf.

Lohrmann, D. (1995) 'Von der östlichen zur westlichen Windmühle'. *Archiv für Kulturgeschichte* 77(1): 1–30 (10f).

Longcore, T. and C. Rich (2004) 'Ecological light pollution'. *Frontiers in Ecology and the Environment* 2(4): 191–8.

Longhurst, J. W. S., S. J. Lindley, A. F. R. Watson and D. E. Conlan (1996) 'The introduction of local air quality management in the United Kingdom: a review and theoretical framework'. *Atmospheric Environment* 30(23): 3975–85.

Longhurst, J. W. S., C. I. Beattie, T. Chatterton, E. T. Hayes, N. S. Leksmono and N. K. Woodfield (2006) 'Local air quality management as a risk management process: assessing, managing and remediating the risk of exceeding an air quality objective in Great Britain'. *Environment International* 32(8): 934–47.

Longhurst, J. W. S., J. G. Irwin, T. J. Chatterton, E. T. Hayes, N. S. Leksmono and J. K. Symons (2009) 'The development of effects-based air quality management regimes'. *Atmospheric Environment* 43(1): 64–78.

Maitland, F. W. (1886) 'The mystery of seisin'. *Law Quarterly Review* 481.

Malakoff, D. (2001) 'Faulty towers'. *Audubon* 103(5): 78–83.

Mann, M. E., J. D. Woodruff, J. P. Donnelly and Z. Zhang (2009) 'Atlantic hurricanes and climate over the past 1,500 years'. *Nature* 460: 880–83.

Middleton, N. J. and A. S. Goudie (2001) 'Saharan dust: sources and trajectories'. *Transactions of the Institute of British Geographers, New Series* 26: 165–81.

Millennium Ecosystem Assessment (2005a) *Ecosystems and Human Well-being.* Washington, DC: Island Press.

— (2005b) *Ecosystems And Human Well-Being: Wetlands and water synthesis.* Washington, DC: World Resources Institute.

Mills, G. and H. Harmens (eds) (2011) *Ozone Pollution: A hidden threat to food security. Report prepared by the ICP Vegetation.* Bangor: Centre for Ecology and Hydrology. icpvegetation.ceh.ac.uk/publications/documents/ozoneandfoodsecurity-ICPVegetationreport%202011-published.pdf.

Mitra, S., R. Wassmann and P. L. G. Vlek (2005) 'An appraisal of global wetland area and its organic carbon stock'. *Current Science* 88: 25–35.

Moolgavkar, S. H. and E. G. Luebeck (1996) 'A critical review of the evidence on particulate air pollution and mortality'. *Epidemiology* 7(4): 420–8.

Moore, M. (2014) 'China's "airpocalypse" kills 350,000 to 500,000 each year'. *The Telegraph*, 7 January. www.telegraph.co.uk/news/worldnews/asia/china/10555816/Chinas-airpocalypse-kills-350000-to-500000-each-year.html.

Morrisette, P. M. (1989) 'The evolution of policy responses to stratospheric ozone depletion'. *Natural Resources Journal* 29: 793–820.

Najam, A. (2001) 'Future directions: the case for a "law of the atmosphere"'. *Atmospheric Environment* 34: 4047–9.

National Academy of Sciences (1976) *Halocarbons: Effects on stratospheric ozone*. Washington, DC: National Academy of Sciences.

Newman, P. A., E. R. Nash, S. R. Kawa, S. A. Montzka and S. M. Schauffler (2006) 'When will the Antarctic ozone hole recover?' *Geophysical Research Letters* 33(12): L12814.

NSCA (2000) *2001 Pollution Handbook*. Brighton: National Society for Clean Air (NSCA).

Nuttall, P. M., A. G. Boon and M. R. Rowell (1997) *Review of the Design and Management of Constructed Wetlands*. CIRIA Report 180. London: Construction Industry Research and Information Association.

OECD (2010) *Paying for Biodiversity: Enhancing the cost-effectiveness of payments for ecosystem services*. Paris: Organisation for Economic Co-operation and Development (OECD) Publishing. doi: 10.1787/9789264090279-en.

— (2013) *OECD Compendium of Agri-environmental Indicators*. Paris: Organisation for Economic Co-operation and Development (OECD) Publishing, pp. 1–185.

Pantages, E. and C. Dulac (2000) 'A novel family of candidate pheromone receptors in mammals'. *Neuron* 28(3): 835–45. doi: 10.1016/S0896-6273(00)00157-4.

Perrot-Maître, D. (2006) *The Vittel Payments for Ecosystem Services: A 'perfect' PES case?* London: International Institute for Environment and Development.

Perry, G., B. W. Buchanan, R. N. Fisher, M. Salmon and S. E. Wise (2008) 'Effects of artificial night lighting on amphibians and reptiles in urban environments'. In J. C. Mitchell, R. E. Jung Brown and B. Bartholomew (eds) *Urban Herpetology*. Salt Lake City, UT: Society for the Study of Amphibians and Reptiles, pp. 239–56.

Perry, K. and J. Edgar (2014) 'Sahara dust storm prompts "serious" health warning for asthmatics'. *The Telegraph*, 2 April. www.telegraph.co.uk/earth/environment/10739019/Sahara-dust-storm-prompts-serious-health-warning-for-asthmatics.html.

Pfaff, A., S. Kerr, L. Lipper, R. Cavatassi, B. Davis, J. Hendy and A. Sanchez (2007) 'Will buying tropical forest carbon benefit the poor? Evidence from Costa Rica'. *Land Use Policy* 24(3): 600–10.

Pijnenburg, L., M. Camps and G. Jongmans-Liedekerken (1991) *Looking Closer at Assimilation Lighting*. Venlo: GGD Limburg-Noord.

Plato (2012) *Trial and Death of Socrates*. Translated by Benjamin Jowett. Barnes & Noble Library of Essential Reading. New York, NY: Barnes & Noble, Inc.

Plutchik, R. (2002) *Emotions and Life: Perspectives from psychology, biology, and evolution.* Washington, DC: American Psychological Association.

Plutchik, R. and H. Kellerman (1980) *Emotion: Theory, research, and experience. Vol. 1. Theories of emotion.* New York, NY: Academic Press.

Porter, S. (2011) 'The rise of predators'. *Geology* 39(6): 607–8.

Pugh, T. A. M., A. M. MacKenzie, J. D. Whyatt and C. N. Hewitt (2012) 'Effectiveness of green infrastructure for improvement of air quality in urban street canyons'. *Environmental Science and Technology* 46: 7692–9. doi: 10.1021/es300826w.

Radle, A. L. (2007) 'The effect of noise on wildlife: a literature review'. World Forum for Acoustic Ecology Online Reader. www.wfae.proscenia.net/library/articles/radle_effect_noise_wildlife.pdf.

Rafaj, P., W. Schöpp, P. Russ et al. (2013) 'Co-benefits of post-2012 global climate mitigation policies'. *Mitigation and Adaptation Strategies for Global Change* 18: 801–24. doi: 10.1007/s11027-012-9390-6.

Rahaman, M. and O. Varis (2005) 'Integrated water resource management: evolution, prospects and future challenges'. *Sustainability: Science, Practice & Policy* 1(1): 15–21.

Ramsar Convention (2002) 'The 8th meeting of the Conference of the Contracting Parties to the Ramsar Convention: The Ramsar Convention on Wetlands'. Gland, Switzerland: Ramsar Convention Secretariat. www.ramsar.org/library.

Randall, D. K. (2007) 'Maybe only god can make a tree, but only people can put a price on it'. *New York Times*, 18 April. www.nytimes.com/2007/04/18/nyregion/18trees.html?module=Search&mabReward=relbias%3As%2C%7B%222%22%3A%22RI%3A13%22%7D&_r=0.

Razzouk, A. W. (2013) 'Fracking: American dream, Chinese pipe-dream, global nightmare'. *Independent*, 13 August. www.independent.co.uk/voices/comment/fracking-american-dream-chinese-pipedream-global-nightmare-8759128.html.

RCEP (2000) *22nd report: Energy – The changing climate*. London: Royal Commission on Environmental Pollution (RCEP).

Robèrt, K.-H. (2002) *The Natural Step Story: Seeding a quiet revolution.* Gabriola Island, Canada: New Society Publishers.

Roberts, M. (2014) 'EU air pollution target "still too high" for heart health'. *BBC News Health*, 22 January. www.bbc.co.uk/news/health-25827304.

Rowland, K. (2012) 'Recall of the wild'. *Green Futures* 86, October: 26–8.

Royal Society of Chemistry (1996) *Simple Guide on Management and Control of Wastes*. London: Royal Society of Chemistry.

Salmon, M. (2003) 'Artificial night lighting and sea turtles'. *Biologist* 50: 163–8.

Samways, M. (1999) 'Translocating fauna to foreign lands: here comes the Homogenocene'. *Journal of Insect Conservation* 3(2): 65–6.

Santos, G., H. Maoh, D. Potoglou et al. (2013) 'Factors influencing modal split of commuting journeys in medium-size European cities'. *Journal of Transport Geography* 30: 127–37.

Sarkar, A. and S. B. Agrawal (2010)

'Elevated ozone and two modern wheat cultivars: an assessment of dose dependent sensitivity with respect to growth, reproductive and yield parameters'. *Environmental and Experimental Botany* 69(3): 328–37.

Sax, J. (1970) 'The public trust doctrine'. *Michigan Law Review* 68(3): 471–566. doi: 10.2307/1287556.

Schernhammer, E. S. and K. Schulmeister (2004) 'Melatonin and cancer risk: does light at night compromise physiologic cancer protection by lowering serum melatonin levels'. *British Journal of Cancer* 90(5): 941–3.

Schernhammer, E. S., F. Laden, F. E. Speizer, W. C. Willett, D. J. Hunter, I. Kawachi and G. A. Colditz (2001) 'Rotating night shifts and risk of breast cancer in women participating in the nurses' health study'. *Journal of the National Cancer Institute* 93(20): 1563–8.

Scholz, M., R. Harrington, P. Carroll and A. Mustafa (2007) 'The Integrated Constructed Wetlands (ICW) concept'. *Wetlands* 27(2): 337–54.

Shapley, H. (1967) *Beyond the Observatory*. New York, NY: Charles Scribner's Sons.

Sigel, S., H. Sigel and R. K. O. Sigel (eds) (2008) *Biomineralization: From nature to application. Metal ions in life sciences*. Chichester: John Wiley & Sons.

Small, C. and J. E. Cohen (2004) 'Continental physiography, climate, and the global distribution of human population'. *Current Anthropology* 45(2): 269–77.

Smart, J. C. R., K. Hicks, T. Morrissey, A. Heinemeyer, M. A. Sutton and M. Ashmore (2011) 'Applying the ecosystem service concept to air quality management in the UK: a case study for ammonia'. *Environmetrics* 22(5): 649–61. doi: 10.1002/env.1094.

Smith, B., R. L. Richards and W. E. Newton (2004) *Catalysts for Nitrogen Fixation: Nitrogenases, relevant chemical models and commercial processes*. Dordrecht and Boston, MA: Kluwer Academic Publishers.

Smith, S., P. Rowcroft, M. Everard, L. Couldrick, M. Reed, H. Rogers, T. Quick, C. Eves and C. White (2013) *Payments for Ecosystem Services: A best practice guide*. London: Department for Environment, Food and Rural Affairs.

Speth, J. G. (2004) *Red Sky at Morning: America and the crisis of the global environment*. New Haven, CT: Yale University Press.

Staddon, C., T. Appleby and E. Grant (2012) 'The human right to water: geographico-legal perspectives'. In F. Sultana and A. Loftus (eds) *The Right to Water*. London: Earthscan Press.

Stern, N. (2006) *Stern Review on the Economics of Climate Change*. London: HM Treasury. webarchive.nationalarchives.gov.uk/+/http:/www.hm-treasury.gov.uk/sternreview_index.htm.

Stuuta, J.-B., I. Smalley and K. O'Hara-Dhand (2009) 'Aeolian dust in Europe: African sources and European deposits'. *Quaternary International* 198(1–2): 234–45.

Sutton, M. and H. van Grinsven (2011) 'Summary for policy makers'. In M. Sutton et al. (eds) *The European Nitrogen Assessment: Sources, effects and policy perspectives*. Cambridge: Cambridge University Press, pp. xxiv–xxxiv.

Suzuki, D. and A. McConnell (1997) *The Sacred Balance: Rediscovering*

our place in nature. Vancouver: Greystone Books.
Thornes, J. E. (1999) *John Constable's Skies: A fusion of art and science*. Edgbaston, Birmingham: University of Birmingham Press.
— (2008) 'Cultural climatology and the representation of sky, atmosphere, weather and climate in selected art works of Constable, Monet and Eliasson'. *Geoforum* 39(2): 570–80.
— (2011) 'Atmospheric services: a new framework for the management of climate change'. Atmospheric Services session, International Conference of the Royal Geographical Society, London, 2 September.
Thornes J. E., W. Bloss, S. Bouzarovski, X. Cai, L. Chapman, J. Clark, S. Dessai, S. Du, D. van der Horst, M. Kendall, C. Kidda and S. Randalls (2010) 'Communicating the value of atmospheric services'. *Meteorological Applications* 17: 243–50.
Times of India (2013) 'Air pollution in Mumbai doubles'. *Times of India*, 25 January. articles.timesofindia.indiatimes.com/2013-01-25/mumbai/36547794_1_air-pollution-spm-pollution-levels.
Tropical Forest Group (2011) *REDD+ and the United Nations Framework Convention on Climate Change (UNFCCC): Justification and recommendations for a new REDD+ mechanism*. San Diego, CA: Tropical Forest Group. unfccc.int/resource/docs/2011/smsn/ngo/241.pdf.
Umweltbundesamt (2010) *2050: 100%. Energy Target 2050: 100% renewable electricity supply*. Dessau-Roßlau: Umweltbundesamt. www.umweltbundesamt.de/sites/default/files/medien/publikation/add/3997-0.pdf.
UN (2002) *Report of the World Summit on Sustainable Development: Johannesburg, South Africa, 26 August–4 September 2002*. New York, NY: United Nations (UN).
UNECE (1979) 'Convention on long-range transboundary air pollution'. Geneva: United Nations Economic Commission for Europe (UNECE). www.unece.org/env/lrtap.
— (1998) 'Convention on access to information, public participation in decision-making and access to justice in environmental matters'. Geneva: United Nations Economic Commission for Europe (UNECE). www.unece.org/env/pp/documents/cep43e.pdf.
University of Washington (1998) 'Urban forest values: economic benefits of trees in cities'. Factsheet 3. Washington, DC: University of Washington, Center for Urban Horticulture. www.naturewithin.info/Policy/EconBens-FS3.pdf.
US EPA (undated) 'Myth: ozone depletion occurs only in Antarctica'. US Environmental Protection Agency (EPA) website. www.epa.gov/ozone/science/myths/glob_dep.html.
— (2008) 'Acid rain'. United States Environmental Protection Agency (EPA) website. www.epa.gov/acidrain/.
Verheijen, F. J. (1985) 'Photopollution: artificial light optic spatial control systems fail to cope with. Incidents, causation, remedies'. *Experimental Biology* 44(1): 1–18.
Vitousek, P. M., J. Aber, R. W. Howarth, G. E. Likens, P. A. Matson, D. W. Schindler, W. H. Schlesinger and G. D. Tilman (1997) *Human Alteration of the Global Nitrogen Cycle: Causes and consequences*. Issues in Ecology 7. Washington, DC: Ecological Society of America. cfpub.epa.gov/watertrain/pdf/issue1.pdf.

Voa, Q. T., C. Kuenzerb, Q. M. Voc, F. Moderd and N. Oppelte (2012) 'Review of valuation methods for mangrove ecosystem services'. *Ecological Indicators* 23: 431–46.

Vogler, J. (2001) 'Future directions: the atmosphere as a global commons'. *Atmospheric Environment* 35: 2427–8.

Votsi, N. E. P., E. G. Drakou, A. D. Mazaris, A. S. Kallimanis and J. D. Pantis (2012) 'Distance-based assessment of open country quiet areas in Greece'. *Landscape and Urban Planning* 104: 279–88.

Walker, G. (2007) *An Ocean of Air: A natural history of the atmosphere*. London: Bloomsbury Press.

Waters, R. D., J. Lusardi and S. J. Clarke (2012) *Delivering the Ecosystem Approach on the Ground: An evaluation of the upland ecosystem service pilots*. Research Report NERR046. York: Natural England.publications. naturalengland.org.uk/file/4092101.

Watson, L. (1974) *Supernature: A natural history of the supernatural*. London: Coronet Books.

Webster, B. and R. Pagnamenta (2010) 'UN must investigate warming "bias", says former climate chief'. *Times Online*, 15 February. www.thetimes.co.uk/tto/environment/article2144989.ece.

Weiner, S. and H. A. Lowenstam (1989) *On Biomineralization*. Oxford: Oxford University Press.

WHO (2010) *Dioxins and Their Effects on Human Health*. World Health Authority Fact Sheet No. 225. Geneva: World Health Organization. www.who.int/mediacentre/factsheets/fs225/en/.

Wilford, J. N. (2000) 'Ages-old icecap at North Pole is now liquid, scientists find'. *New York Times*, 19 August.

Willows, R. and A. Hart (2013) 'Nitrate vulnerable zones: a good example of risk-based, targeted regulation?' *Environmental Scientist*, November: 23–6.

WMO (2013) *WMO Greenhouse Gas Bulletin No. 9*. Geneva: World Meteorological Organization (WMO). www.wmo.int/pages/prog/arep/gaw/ghg/documents/GHG_Bulletin_No.9_en.pdf.

Woods-Ballard, B., R. Kellagher, P. Martin, C. Jefferies, R. Bray and P. Shaffer (2007) *The SuDS Manual*. CIRIA Report C697. London: Construction Industry Research and Information Association.

Wünscher, T., S. Engel and S. Wunder (2006) 'Payments for environmental services in Costa Rica: increasing efficiency through spatial differentiation'. *Quarterly Journal of International Agriculture* 45(4): 319–37.

Yandle, Y. (2006) 'Property rights in two states of nature: a discussion of evolving rights'. Published in French as 'l'évolution des droits de propriété'. In M. Falque, H. Lamotte and J.-F. Saglio (eds) *Les Ressources Foncières: Droits de Propriété, Économie et Environnement*. Proceedings of the VIth International Conference, International Center for Research on Environmental Issues, Paul Cézanne University, Aix-en-Provence, 26–28 June.

Zhu, Y. and R. E. Newell (1994) 'Atmospheric rivers and bombs'. *Geophysics Research Letters* 21(18): 1999–2002.

— (1998) 'A proposed algorithm for moisture fluxes from atmospheric rivers'. *Monthly Weather Review* 126(3): 725–35.

Index